Springer Series in Statistics

Springer
New York
Berlin
Heidelberg
Barcelona
Budapest
Hong Kong
London
Milan
Paris
Santa Clara
Singapore
Tokyo

Springer Series in Statistics

Andersen/Borgan/Gill/Keiding: Statistical Models Based on Counting Processes.
Andrews/Herzberg: Data: A Collection of Problems from Many Fields for the Student and Research Worker.
Anscombe: Computing in Statistical Science through APL.
Berger: Statistical Decision Theory and Bayesian Analysis, 2nd edition.
Bolfarine/Zacks: Prediction Theory for Finite Populations.
Borg/Groenen: Modern Multidimensional Scaling: Theory and Applications
Brémaud: Point Processes and Queues: Martingale Dynamics.
Brockwell/Davis: Time Series: Theory and Methods, 2nd edition.
Daley/Vere-Jones: An Introduction to the Theory of Point Processes.
Dzhaparidze: Parameter Estimation and Hypothesis Testing in Spectral Analysis of Stationary Time Series.
Fahrmeir/Tutz: Multivariate Statistical Modelling Based on Generalized Linear Models.
Farrell: Multivariate Calculation.
Federer: Statistical Design and Analysis for Intercropping Experiments.
Fienberg/Hoaglin/Kruskal/Tanur (Eds.): A Statistical Model: Frederick Mosteller's Contributions to Statistics, Science and Public Policy.
Fisher/Sen: The Collected Works of Wassily Hoeffding.
Good: Permutation Tests: A Practical Guide to Resampling Methods for Testing Hypotheses.
Goodman/Kruskal: Measures of Association for Cross Classifications.
Gouriéroux: ARCH Models and Financial Applications.
Grandell: Aspects of Risk Theory.
Haberman: Advanced Statistics, Volume I: Description of Populations.
Hall: The Bootstrap and Edgeworth Expansion.
Härdle: Smoothing Techniques: With Implementation in S.
Hart: Nonparametric Smoothing and Lack-of-Fit Tests.
Hartigan: Bayes Theory.
Heyer: Theory of Statistical Experiments.
Huet/Bouvier/Gruet/Jolivet: Statistical Tools for Nonlinear Regression: A Practical Guide with S-PLUS Examples.
Jolliffe: Principal Component Analysis.
Kolen/Brennan: Test Equating: Methods and Practices.
Kotz/Johnson (Eds.): Breakthroughs in Statistics Volume I.
Kotz/Johnson (Eds.): Breakthroughs in Statistics Volume II.
Kres: Statistical Tables for Multivariate Analysis.
Le Cam: Asymptotic Methods in Statistical Decision Theory.
Le Cam/Yang: Asymptotics in Statistics: Some Basic Concepts.
Longford: Models for Uncertainty in Educational Testing.
Manoukian: Modern Concepts and Theorems of Mathematical Statistics.
Miller, Jr.: Simultaneous Statistical Inference, 2nd edition.
Mosteller/Wallace: Applied Bayesian and Classical Inference: The Case of *The Federalist Papers.*

(continued after index)

J.O. Ramsay B.W. Silverman

Functional Data Analysis

With 98 Figures

Springer

J.O. Ramsay
Department of Psychology
McGill University
Montreal, Quebec H3A 1B1
Canada

B.W. Silverman
Department of Mathematics
University of Bristol
University Walk
Bristol BS8 1TW, United Kingdom

Library of Congress Cataloging-in-Publication Data
Ramsay, J. O. (James O.)
Functional data analysis / J.O. Ramsay, B.W. Silverman.
p. cm. — (Springer series in statistics)
Includes bibliographical references (p. -) and index.
ISBN 0-387-94956-9 (hc : alk. paper)
1. Multivariate analysis. I. Silverman, B. W., 1952–
II. Title. III. Series.
QA278.R36 1997
519.5—dc21 96-54729

Printed on acid-free paper.

Production managed by Terry Kornak; manufacturing supervised by Joe Quatela.
Camera-ready copy prepared from the authors' LaTeX files.
Printed and bound by Maple-Vail Book Manufacturing Group, York, PA.
Printed in the United States of America.

9 8 7 6 5 4 3 2 1

ISBN 0-387-94956-9 Springer-Verlag New York Berlin Heidelberg SPIN 10557897

Preface

This book is a snapshot of a highly social and, therefore, decidedly unpredictable process. The combined personal view of functional data analysis that it presents has emerged over a number of years of research and contact, and has been greatly nourished by delightful collaborations with many friends. We hope that readers will enjoy the book as much as we have enjoyed writing it, whether they are our colleagues as researchers or applied data analysts reading the book as a research monograph, or students using it as a course text.

After some introductory material in Chapters 1 and 2, we discuss aspects of smoothing methods in Chapters 3 and 4. Chapter 5 deals with curve registration, the alignment of common characteristics of a sample of curves. In the next three chapters, we turn to functional principal components analysis, one of our main exploratory techniques. In Chapters 9 to 11, problems that involve covariate information are considered in various functional versions of the linear model. The functional analogue of canonical correlation analysis in Chapter 12 explores the relationship between two functional variables treating them symmetrically. Chapters 13 to 15 develop specifically functional methods that exploit derivative information and the use of linear differential operators. Our final chapter provides historical remarks and some pointers to possible future developments.

Data arising in real applications are used throughout for both motivation and illustration, showing how functional approaches allow us to see new things, especially by exploiting the smoothness of the processes generating the data. The data sets exemplify the wide scope of functional data analysis; they are drawn from growth analysis, meteorology, biomechanics, equine science, economics, and medicine. Some of these data sets, together with the software we have used to analyse them, are available from a world-wide web site described at the end of Chapter 1.

Many people have been of great help to us in our work on this book. In particular we would like to thank Michal Abrahamowicz, Philippe Besse, Darrell Bock, Jeremy Burn, Catherine Dalzell, Shelly Feran, Randy Flanagan, Rowena Fowler, Theo Gasser, Mary Gauthier, Vince Gracco, David Green, Nancy Heckman, Anouk Hoedeman, Steve Hunka, Iain Johnstone, Alois Kneip, Wojtek Krzanowski, Xiaochun Li, Duncan Murdoch, Kevin Munhall, Guy Nason, Richard Olshen, David Ostry, Tim Ramsay, John Rice, Xiaohui Wang and Alan Wilson. We gratefully acknowledge financial support from the Natural Science and Engineering Research Council of Canada, the National Science Foundation and the National Institute of Health of the USA, and the British Engineering and Physical Sciences Research Council. Above all, our thanks are due to the Royal Statistical Society Research Section; our first contact was at a discussion meeting where one of us read a paper and the other proposed the vote of thanks, not always an occasion that leads to a meeting of minds!

November 1996

Jim Ramsay
Bernard Silverman

Contents

1
Introduction

1.1 What are functional data?

Figure 1.1 provides a prototype for the type of data that we shall consider. It shows the heights of 10 Swiss boys measured at a set of 29 ages in the Zurich Longitudinal Growth Study (Falkner, 1960). The ages are not equally spaced; there are annual measurements from two to 10 years, followed by biannually measured heights. Although great care was taken in the measurement process, there is an uncertainty or noise in height values with a standard deviation of about 5 mm. Even though each record involves only 29 discrete values, these values reflect a smooth variation in height that could be assessed, in principle, as often as desired, and is therefore a height *function*. Thus, the data consist of a sample of 10 *functional* observations $\texttt{Height}_i(t)$.

Moreover, there are features in this data too subtle to see in this type of plot. For example, Gasser et al. (1984) detected a growth spurt prior to puberty using some delicate and powerful data analysis tools. Figure 1.2 displays the acceleration curves $D^2\texttt{Height}_i$ estimated from these data by Ramsay, Bock and Gasser (1995) using a technique discussed in Chapter 4. We use the notation D for differentiation, as in

$$D^2\texttt{Height} = \frac{d^2\texttt{Height}}{dt^2}.$$

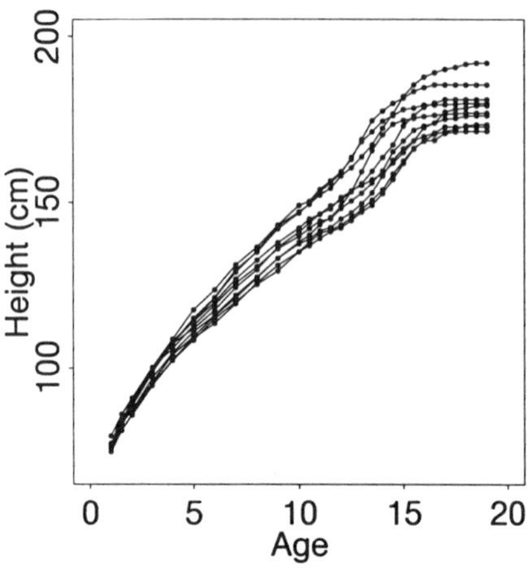

FIGURE 1.1. The heights of 10 Swiss boys measured at 29 ages. The points indicate the unequally spaced ages of measurement.

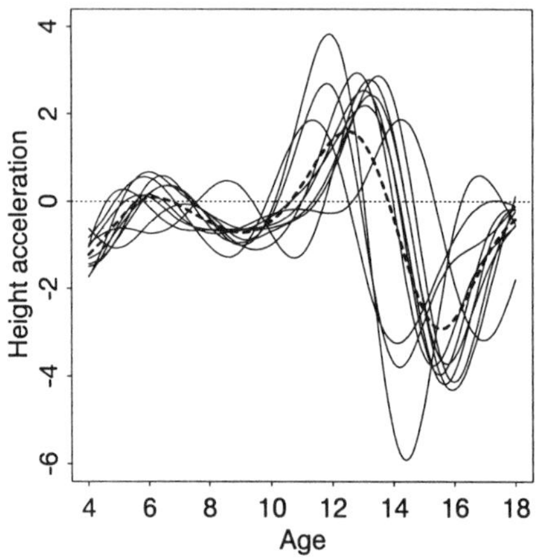

FIGURE 1.2. The estimated accelerations of height for 10 Swiss boys, measured in centimetres per year per year. The heavy dashed line is the cross-sectional mean, and is a rather poor summary of the curves.

In Figure 1.2 the pubertal growth spurt shows up as a pulse of strong positive acceleration followed by sharp negative deceleration. But most records also show a bump at around six years that is termed the mid-spurt. We conclude that the variation from curve to curve can also be profitably explained at the level of certain derivatives. The fact that derivatives are of interest is further reason to think of the records as functions, rather than vectors of observations in discrete time.

The ages themselves must also play an explicit role in our analysis, because they are not equally spaced. Although it might be mildly interesting to correlate heights at ages 9, 10 and 10.5, this would not take account of the fact that we fully expect the correlation for two ages separated by only half a year to be higher than that for a separation of one year. Indeed, although in this particular example the ages at which the observations are taken are nominally the same for each boy, there is no real need for this to be so; in general, the points at which the functions are observed may well vary from one record to another.

The replication of these height curves invites an exploration of the ways in which the curves vary. This is potentially complex. For example, the rapid growth during puberty is visible in all curves, but both the timing and the intensity of pubertal growth differ from boy to boy. Some type of principal components analysis would undoubtedly be helpful, but we must adapt the procedure to take account of the unequal age spacing and the smoothness of the underlying height functions. One objective might be to separate variation in timing of significant growth events, such as the pubertal growth spurt, from variation in the intensity of growth.

1.2 Some functional data analyses

Data in many different fields come to us through a process naturally described as functional. To turn to a completely different context, consider Figure 1.3, where the mean monthly temperatures for four Canadian weather stations are plotted. It also shows estimates of the corresponding smooth temperature functions presumed to generate the observations. Montreal, with the warmest summer temperature, has a temperature pattern that appears to be nicely sinusoidal. Edmonton, with the next warmest summer temperature, seems to have some distinctive departures from sinusoidal variation that call for some explanation. The marine climate of Prince Rupert is evident in the small amount of annual variation in temperature, and Resolute has bitterly cold but strongly sinusoidal temperature.

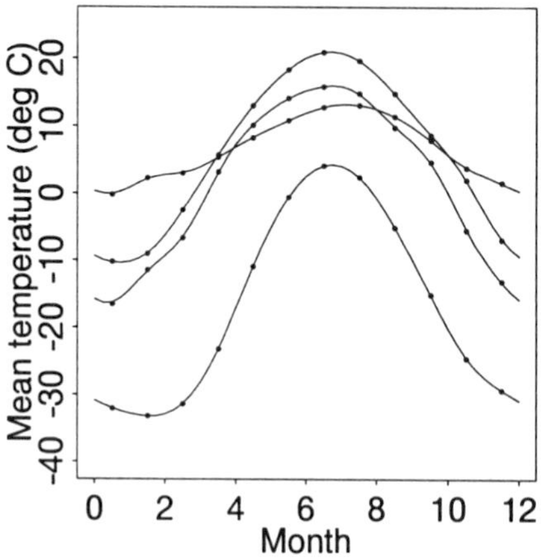

FIGURE 1.3. Mean monthly temperatures for the Canadian weather stations. In descending order of the temperatures at the start of the year, the stations are Prince Rupert, Montreal, Edmonton, and Resolute.

One expects temperature to be primarily sinusoidal in character, and certainly periodic over the annual cycle. There is some variation in phase, because the coldest day of the year seems to be later in Edmonton and Resolute. Consequently, a model of the form

$$\texttt{Temp}_i(t) \approx c_{i1} + c_{i2}\sin(\pi t/6) + c_{i3}\cos(\pi t/6) \tag{1.1}$$

should do rather nicely for these data, where $\texttt{Temp}_i$ is the temperature function for the ith weather station, and (c_{i1}, c_{i2}, c_{i3}) is a vector of three parameters associated with that station.

In fact, there are clear departures from sinusoidal or simple harmonic behaviour. One way to see this is to compute the function

$$L\texttt{Temp} = (\pi/6)^2 D\texttt{Temp} + D^3\texttt{Temp}.$$

As we have already noted in Section 1.1, the notation $D^m\texttt{Temp}$ means "take the mth derivative of function Temp," and the notation $L\texttt{Temp}$ stands for the function which results from applying the linear differential operator $L = (\pi/6)^2 D + D^3$ to the function Temp. The resulting function, $L\texttt{Temp}$, is often called a *forcing function.* Now, if a temperature function is truly sinusoidal, then $L\texttt{Temp}$ should be exactly zero, as it would be for any function of the form (1.1). That is, it would satisfy the *homogeneous differential equation* $L\texttt{Temp} = 0$.

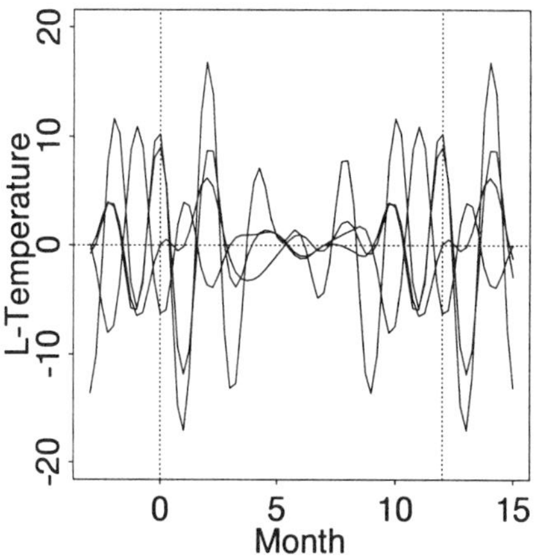

FIGURE 1.4. The result of applying the differential operator $L = (\pi/6)^2 D + D^3$ to the estimated temperature functions in Figure 1.3. If the variation in temperature were purely sinusoidal, these curves would be exactly zero.

But Figure 1.4 indicates that the functions $L\mathtt{Temp}_i$ display systematic features that are especially strong in the spring and autumn months. Put another way, temperature at a particular weather station can be described as the solution of the *nonhomogeneous* differential equation $L\mathtt{Temp} = u$, where the forcing function u can be viewed as input from outside of the system, or an exogenous influence. Meteorologists suggest, for example, that these spring and autumn effects are partly due to the change in the reflectance of land when snow or ice melts, and this would be consistent with the fact that the least sinusoidal records are associated with continental stations well separated from large bodies of water.

Here, the point is that we may often find it interesting to remove effects of a simple character by applying a differential operator, rather than simply subtracting them. This exploits the intrinsic smoothness in the process, and long experience in the natural and engineering sciences suggests that this may get closer to the underlying driving forces at work than just adding and subtracting effects, as one routinely does in multivariate data analysis.

Functional data are often multivariate in a different sense. Our third example is in Figure 1.5. The Motion Analysis Laboratory at Children's

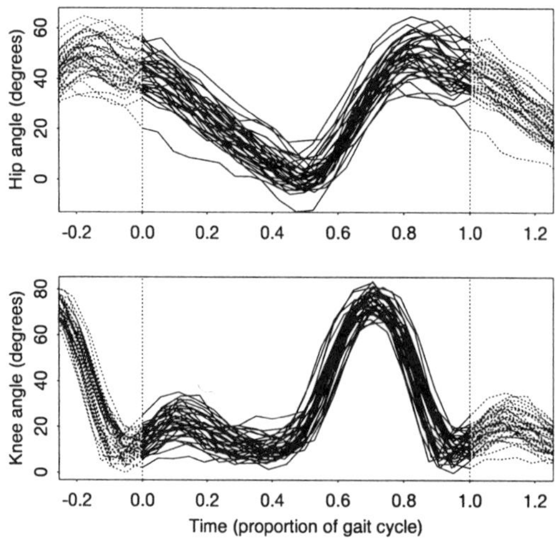

FIGURE 1.5. The angles in the sagittal plane formed by the hip and by the knee as 39 children go through a gait cycle. The interval $[0, 1]$ is a single cycle, and the dotted curves show the periodic extension of the data beyond either end of the cycle.

Hospital, San Diego, collected these data, which consist of the angles formed by the hip and knee of each of 39 children over each child's gait cycle. See Olshen et al. (1989) for full details. Time is measured in terms of the individual gait cycle, so that every curve is given for values of t in $[0, 1]$. The cycle begins and ends at the point where the heel of the limb under observation strikes the ground. Both sets of functions are periodic, and are plotted as dotted curves somewhat beyond the interval for clarity. We see that the knee shows a two-phase process, while the hip motion is single-phase. What is harder to see is how the two joints interact; of course the figure does not indicate which hip curve is paired with which knee curve, and among many other things this example demonstrates the need for graphical ingenuity in functional data analysis.

Figure 1.6 shows the gait cycle for a single child by plotting knee angle against hip angle as time progresses round the cycle. The periodic nature of the process implies that this forms a closed curve. Also shown for reference purposes is the same relationship for the average across the 39 children. Now we see an interesting feature: a cusp occurring

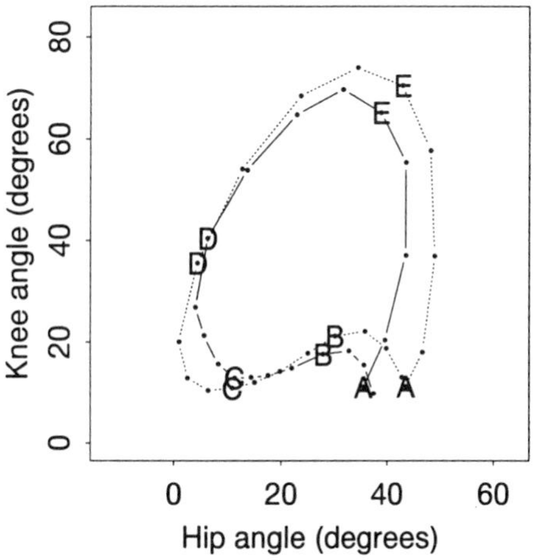

FIGURE 1.6. Solid line: The angles in the sagittal plane formed by the hip and by the knee for a single child plotted against each other. Dotted line: The corresponding plot for the average across children. The points indicate 20 equally spaced time points in the gait cycle, and the letters are plotted at intervals of one-fifth of the cycle, with A marking the heel strike.

at the heel strike. The angular velocity is clearly visible in terms of the spacing between numbers, and it varies considerably as the cycle proceeds. The child whose gait is represented by the solid curve differs from the average in two principal ways. First, the portion of the gait pattern in the C–D part of the cycle shows an exaggeration of movement relative to the average, and second, in the part of the cycle where the hip is most bent, the amount by which the hip is bent is markedly less than average; interestingly, this is not accompanied by any strong effect on the knee angle. The overall shape of the cycle for the particular child is rather different from the average. The exploration of variability in these functional data must focus on features such as these.

Finally, in this introduction to types of functional data, we must not forget that they may come to our attention as full-blown functions, so that each record may consist of functions observed, for all practical purposes, everywhere. Sophisticated on-line sensing and monitoring equipment is now routinely used in research in medicine, seismology, meteorology, physiology, and many other fields.

1.3 The goals of functional data analysis

The goals of functional data analysis are essentially the same as those of any other branch of statistics. They include the following aims:

- to represent the data in ways that aid further analysis
- to display the data so as to highlight various characteristics
- to study important sources of pattern and variation among the data
- to explain variation in an outcome or dependent variable by using input or independent variable information
- to compare two or more sets of data with respect to certain types of variation, where two sets of data can contain different sets of replicates of the same functions, or different functions for a common set of replicates.

Subsequent chapters explore each of these themes, and they are introduced only briefly here.

Each of these activities can be conducted with techniques appropriate to certain goals. Another way to characterize the strategy in a data analysis is as *exploratory*, *confirmatory*, or *predictive*. In exploratory mode, the questions put to the data tend to be rather open-ended in the sense that one expects the right technique to reveal new and interesting aspects of the data, as well as to shed to light on known and obvious features. Exploratory investigations tend to consider only the data at hand, with less concern for statements about larger issues such as characteristics of populations or events not observed in the data. Confirmatory analyses, on the other hand, tend to be inferential and to be determined by specific questions about the data. Some type of structure is assumed to be present in the data, and one wants to know whether certain specific statements or hypotheses can be considered confirmed by the data. The dividing line between exploratory and confirmatory analyses tends to be the extent to which probability theory is used, in the sense that most confirmatory analyses are summarized by one or more probability statements. Predictive studies are somewhat less common, and focus on using the data at hand to make a statement about unobserved states, such as the future.

This book develops techniques that are mainly exploratory in nature. The theory of probability associated with functional events is generally regarded as an advanced topic, and might be seen by many readers as inaccessibly technical. Nevertheless, some simple hypothesis tests are certainly possible, and are mentioned here and there. In general, prediction is beyond our scope.

1.4 First steps in a functional data analysis

1.4.1 Data representation: smoothing and interpolation

Assuming that a functional datum for replication i arrives as a set of discrete measured values, $y_{i1}, \ldots, y_{in}$, the first task is to convert these values to a function x_i with values $x_i(t)$ computable for any desired argument value t. If the discrete values are assumed to be errorless, then the process is *interpolation*, but if they have some observational error that needs removing, then the conversion from discrete data to functions may involve *smoothing*. Chapters 3 and 4 offer a survey of these procedures. The *roughness penalty* smoothing method discussed in Chapter 4 will be used much more broadly in many contexts throughout the book, not merely for the purpose of estimating a function from a set of observed values.

The gait data in Figure 1.5 were converted to functions by the simplest of interpolation schemes: joining each pair of adjacent observations by a straight line segment. This approach would be inadequate if we require derivative information. However, one might perform a certain amount of smoothing while still respecting the periodicity of the data by fitting a Fourier series to each record: A constant plus three pairs of sine and cosine terms does a reasonable job for these data. The growth data in Figure 1.1 and the temperature data in Figure 1.3 were smoothed using polynomial smoothing splines, and this more sophisticated technique also provides high quality derivative information.

1.4.2 Data registration or feature alignment

Figure 1.7 shows some biomechanical data. The curves in the figure are twenty records of the force exerted on a meter during a brief pinch by the thumb and forefinger. The subject was required to maintain a certain background force on a force meter and then to squeeze the meter aiming at a specified maximum value, returning afterwards to the background level. The purpose of the experiment was to study the neurophysiology of the thumb-forefinger muscle group. The data were collected at the MRC Applied Psychology Unit, Cambridge, by R. Flanagan; see Ramsay, Wang and Flanagan (1995).

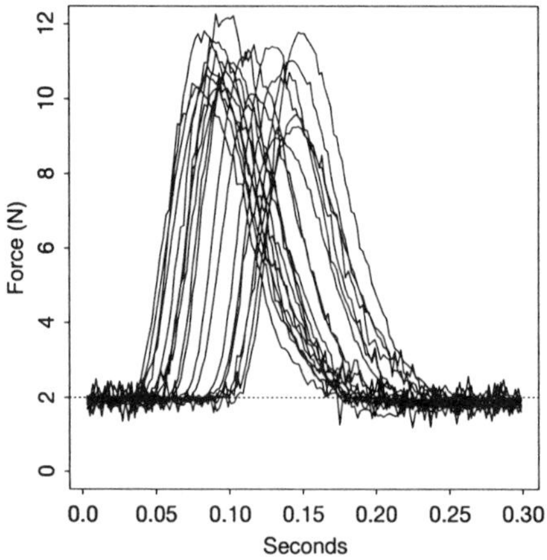

FIGURE 1.7. Twenty recordings of the force exerted by the thumb and forefinger where a constant background force of two newtons was maintained prior to a brief impulse targeted to reach 10 newtons. Force was sampled 500 times per second.

These data illustrate a common problem in functional data analysis. The start of the pinch is located arbitrarily in time, and a first step is to align the records by some shift of the time axis. In Chapter 5 we take up the question of how to estimate this shift, and how to go further if necessary to estimate record-specific linear transformations of the argument, or even nonlinear transformations.

1.4.3 Data display

Displaying the results of a functional data analysis can be a challenge. of the gait data in Figures 1.5 and 1.6, we have already seen that different displays of data can bring out different features of interest, and that the standard plot of $x(t)$ against t is not necessarily the most informative. It is impossible to be prescriptive about the best type of plot for a given set of data or procedure, but we shall give illustrations of various ways of plotting the results. These are intended to stimulate the reader's imagination rather than to lay down rigid rules.

The methods we use do not include such sophistications as three (or more) dimensional displays produced by rotating a structure in more than two dimensions. Instead, we have confined ourselves to plots that can easily be produced in the statistical package S-PLUS (Statistical Sciences, 1995), which we have used as the platform for virtually all our analyses. More sophisticated displays can be informative, but they are inaccessible to printed media. Nevertheless, the use of such techniques in functional data display and analysis is an interesting topic for research.

1.5 Summary statistics for functional data

1.5.1 Mean and variance functions

The classical summary statistics for univariate data familiar to students in introductory statistics classes apply equally to functional data. The mean function with values

$$\bar{x}(t) = N^{-1} \sum_{i=1}^{N} x_i(t)$$

is the average of the functions pointwise across replications. Similarly the variance function $\mathtt{var}$ has values

$$\mathtt{var}_X(t) = (N-1)^{-1} \sum_{i=1}^{N} [x_i(t) - \bar{x}(t)]^2,$$

and the standard deviation function is the square root of the variance function.

Figure 1.8 displays the mean and standard deviation functions for the aligned pinch force data. We see that the mean force looks remarkably like a number of probability density functions well known to statisticians, and in fact the relationship to the lognormal distribution has been explored by Ramsay, Wang and Flanagan (1995). The standard deviation of force seems to be about 8% of the mean force over most of the range of the data.

1.5.2 Covariance and correlation functions

The *covariance function* summarizes the dependence of records across different argument values, and is computed for all t_1 and t_2 by

$$\mathtt{cov}_X(t_1, t_2) = (N-1)^{-1} \sum_{i=1}^{N} \{x_i(t_1) - \bar{x}(t_1)\}\{x_i(t_2) - \bar{x}(t_2)\}.$$

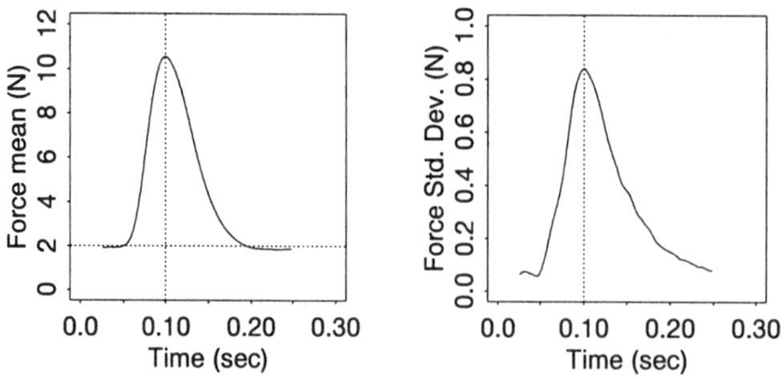

FIGURE 1.8. The mean and standard deviation functions for the 20 pinch force observations in Figure 1.7 after they were aligned or registered.

The associated *correlation function* is

$$\mathrm{corr}_X(t_1, t_2) = \frac{\mathrm{cov}_X(t_1, t_2)}{\sqrt{\mathrm{var}_X(t_1)\mathrm{var}_X(t_2)}}.$$

These are the functional analogues of the variance-covariance and correlation matrices, respectively, in multivariate data analysis.

Figure 1.9 displays the correlation function of the pinch force data, both as a surface over the plane of possible pairs of times (t_1, t_2) and also as a set of level contours.

Our experience with perspective and contour displays of correlation suggests that not everyone encountering them for the first time finds them easy to understand. Here is one strategy: The diagonal running from lower left to upper right in the contour or from front to back in the perspective plot of the surface contains the unit values that are the correlations between identical or very close time values. Directions perpendicular to this ridge of unit correlation indicate how rapidly the correlation falls off as two argument values separate. For example, one might locate a position along the unit ridge associated with argument value t, and then moving perpendicularly from this point shows what happens to the correlation between values at time pair $(t - \delta, t + \delta)$ as the perpendicular distance δ increases. In the case of the pinch force data, we note that the correlation falls off slowly for values on either side of the time 0.1 of maximum force, but declines much more rapidly in the periods before and after the impulse. This suggests a two-phase system, with fairly erratic uncoupled forces in the constant background force phase, but with tightly connected forces during the

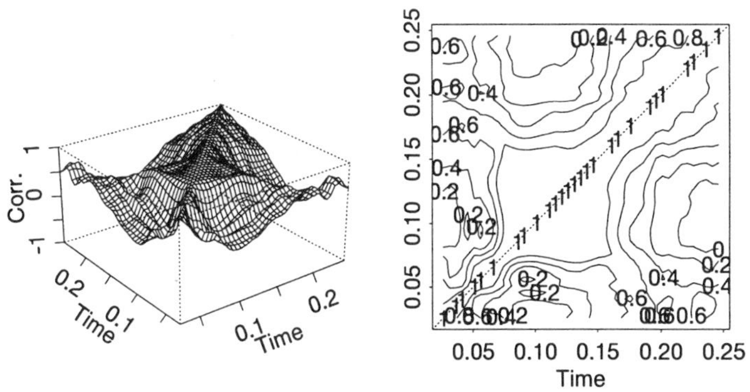

FIGURE 1.9. The left panel is a perspective plot of the bivariate correlation function values $r(t_1, t_2)$ for the pinch force data. The right panel shows the same surface by contour plotting. Time is measured in seconds.

actual impulse. In fact, it is common to observe low correlations or rapid fall-off when a system is in a resting or ballistic state free from any outside input, but to show strong correlations, either positive and negative, when exogenous influences apply.

1.5.3 *Cross-covariance and cross-correlation functions*

In the case of the gait data discussed in Section 1.2, we had both hip and knee angles measured through time. In general, if we have pairs of observed functions (x_i, y_i), the way in which these depend on one another can be quantified by the *cross-covariance* function

$$\text{cov}_{X,Y}(t_1, t_2) = (N-1)^{-1} \sum_{i=1}^{N} \{x_i(t_1) - \bar{x}(t_1)\}\{y_i(t_2) - \bar{y}(t_2)\}.$$

or the *cross-correlation* function

$$\text{corr}_{X,Y}(t_1, t_2) = \frac{\text{cov}_{X,Y}(t_1, t_2)}{\sqrt{\text{var}_X(t_1)\text{var}_Y(t_2)}}.$$

Figure 1.10 displays the correlation and cross-correlation functions for the gait data. In each of the four panels, t_1 is plotted along the horizontal axis and t_2 along the vertical axis. The top left panel shows a contour plot of the correlation function $\text{corr}_{\text{Hip}}(t_1, t_2)$ for the hip angles alone, and the bottom right panel shows the corresponding plot for the knee angles. The cross-correlation functions $\text{corr}_{\text{Hip,Knee}}$

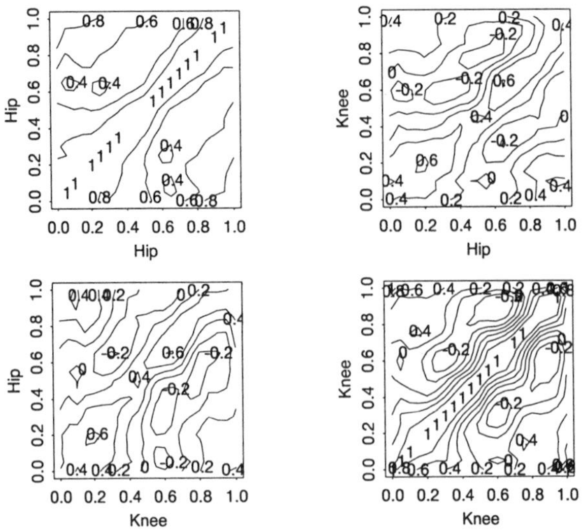

FIGURE 1.10. Contour plots of the correlation and cross-correlation functions for the gait data. In each panel t_1 is plotted on one axis and t_2 on the other; the legends indicate which observations are being correlated against each other.

and $\mathtt{corr}_{\mathtt{Knee,Hip}}$ are plotted in the top right and bottom left panels respectively; since, in general, $\mathtt{corr}_{X,Y}(t_1, t_2) = \mathtt{corr}_{Y,X}(t_2, t_1)$, these are transposes of one another, in that each is the reflection of the other about the main diagonal $t_1 = t_2$. Note that each axis is labelled by the generic name of relevant data function, Hip or Knee, rather than by the argument value t_1 or t_2.

In this figure, different patterns of variability are demonstrated by the individual correlation functions $\mathtt{corr}_{\mathtt{Hip}}$ and $\mathtt{corr}_{\mathtt{Knee}}$ for the hip and knee angles considered separately. The hips show positive correlation throughout, so that if the hip angle is larger than average at one point in the cycle it will have a tendency to be larger than average everywhere. The contours on this plot are more or less parallel to the main diagonal, implying that the correlation is approximately a function of $t_1 - t_2$ and that the variation of the hip angles can be considered as an approximately stationary process.

On the other hand, the knee angles show behaviour that is clearly nonstationary; the correlation between the angle at time 0.0 and time 0.3 is about 0.4, while that between times 0.3 and 0.6 is actually negative. In the middle of the cycle the correlation falls away rapidy as one moves away from the main diagonal, while at the ends of the cycle there is much longer range correlation. The hip angles show a

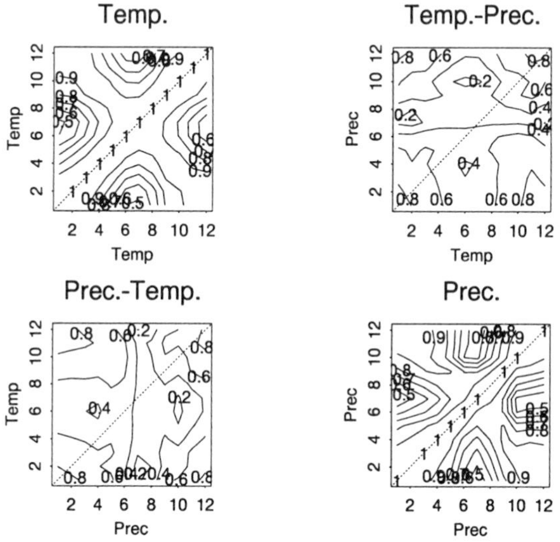

FIGURE 1.11. Contour plots of the correlation and cross-correlation functions for 35 Canadian weather stations for temperature and log precipitation. The cross-correlation functions are those in the upper right and lower left panels.

slight, but much less marked, departure from stationarity of the same kind. These features may be related to the greater effect on the knee of external factors such as the heel strike and the associated weight placed on the joint, whereas the hip acts under much more even muscular control throughout the cycle.

The ridge along the main diagonal of the cross-correlation plots indicates that $\mathtt{Hip}(t_1)$ and $\mathtt{Knee}(t_2)$ are most strongly correlated when t_1 and t_2 are approximately equal, though the main ridge shows a slight reverse S shape (in the orientation of the top right panel). The analysis developed in Chapter 12 will elucidate the delays in the dependence of one joint on the other. Apart from this, there are differences in the way that the cross-correlations behave at different points of the cycle, but the cross-correlation function does not make it clear what these mean in terms of dependence between the functions.

Another example is provided by the Canadian weather data. Contour plotting in Figure 1.11 shows the correlation functions between temperature and log precipitation based on monthly data. The correlation is high for both temperature and precipitation on either side of the midsummer period, so that autumn weather tends to be highly correlated with spring weather. By contrast, winter and summer weather have a weaker correlation of around 0.5. The cross-correlations show that midsummer precipitation has a near zero correlation with temperature at any point in the year, but that midwinter temperature and midwinter precipitation are highly correlated. This is due to the fact that, in continental weather stations, both measures tend to be especially low in midwinter, whereas in marine stations, the tendency is for both temperature and precipitation to be higher.

1.6 Exploring variability in more detail

The examples considered so far give a glimpse of ways in which the variability of a set of functional data is interesting, but there is a need for more detailed and sophisticated ways of investigating variability, and these are a major theme of this book.

1.6.1 Functional principal components analysis

Most sets of data display a small number of dominant or substantial modes of variation, even after subtracting the mean function from each observation. An approach to identifying and exploring these, set out in Chapter 6, is to adapt the classical multivariate procedure of principal components analysis to functional data. In Chapter 7, techniques of smoothing or regularization are incorporated into the functional principal components analysis itself, thereby demonstrating that smoothing methods have a far wider rôle in functional data analysis than merely in the initial step of converting discrete observations to functional form. In Chapter 8, we show that functional principal components analysis can be made more selective and informative by considering specific types of variation in a special way. For example, we shall see that estimating a small shift of time for each temperature record and studying its variation will give a clearer understanding of record-to-record temperature variability.

1.6.2 Functional linear modelling

The classical techniques of linear regression, analysis of variance, and linear modelling all investigate the way in which variability in observed data can be accounted for by other known or observed variables. They can all be placed within the framework of the general linear model

$$y = \mathbf{Z}\beta + \epsilon \tag{1.2}$$

where, in the simplest case, y is typically a vector of observations, β is a parameter vector, $\mathbf{Z}$ is a matrix that defines a linear transformation from parameter space to observation space, and ϵ is an error vector with mean zero. The design matrix $\mathbf{Z}$ incorporates observed covariates or independent variables.

To extend these ideas to the functional context, we retain the basic structure (1.2) but allow more general interpretations of the symbols within it. For example, we might ask of the Canadian weather data:

- If each weather station is broadly categorized as being Atlantic, Pacific, Continental or Arctic, in what way does the geographical category characterize the detailed temperature profile `Temp` and account for the different profiles observed? In Chapter 9 we introduce a functional analysis of variance methodology, where both the parameters and the observations become functions, but the matrix $\mathbf{Z}$ remains the same as in the classical multivariate case.

- Could a temperature record `Temp` be used to predict the logarithm of total annual precipitation? In Chapter 10 we extend the idea of linear regression to the case where the independent variable, or covariate, is a function, but the response variable (log total annual precipitation in this case) is not.

- Can the temperature record `Temp` be used as a predictor of the entire precipitation profile, not merely the total precipitation? This requires a fully functional linear model, where all the terms in the model have more general form than in the classical case. This topic is considered in Chapter 11.

1.6.3 Functional canonical correlation

How do two or more sets of records covary or depend on one another? As we saw in the cross-correlation plots, this is a question to pose for gait data, because relationships between record-to-record variation in hip angle and knee angle seem likely.

The functional linear modelling framework approaches this question by considering one of the sets of functional observations as a covariate and the other as a response variable, but in many cases, such as the gait data, it does not seem reasonable to impose this kind of asymmetry, and we shall develop two rather different methods that treat both sets of variables in an even-handed way. One method, described in Section 6.5, essentially treats the pair $(\mathtt{Hip}_i, \mathtt{Knee}_i)$ as a single vector-valued function, and then extends the functional principal components approach to perform an analysis. Chapter 12 takes another approach, a functional version of canonical correlation analysis, identifying components of variability in each of the two sets of observations which are highly correlated with one another.

For many of the methods we discuss, a naïve approach extending the classical multivariate method will usually give reasonable results, though regularization will often improve these. However, when a linear predictor is based on a functional observation, and also in functional canonical correlation analysis, regularization is not an optional extra but is an intrinsic and necessary part of the analysis; the reasons are discussed in Chapters 10, 11 and 12.

1.7 Using derivatives in functional data analysis

In Section 1.2 we have already had a taste of the ways in which derivatives and linear differential operators are useful in functional data analysis. The use of derivatives is important both in extending the range of simple graphical exploratory methods, and in the development of more detailed methodology. This is a theme that will be explored in much more detail in Chapters 13, 14 and 15, but some preliminary discussion is appropriate here.

1.7.1 Correlation plots for derivatives

The growth curves considered in Section 1.1 illustrated the importance of considering not only the original observations but also some derivative estimated from them, in this case the second derivative or acceleration. Once the acceleration curves are estimated, they can be treated in just the same way as if they were the primary data. For example, Figure 1.12 contains the contour plot of the correlation function for the registered acceleration curves for the growth data. A much tighter coupling of acceleration values is evident in the vicinity of the two growth spurts. For example, there is a negative correlation of about 0.5 between accelerations at the beginning and end of the pubertal growth spurt, whose centre is at 14 years. These spurts tend to have either large or shallow amplitudes, and sharp acceleration is followed by sharp deceleration.

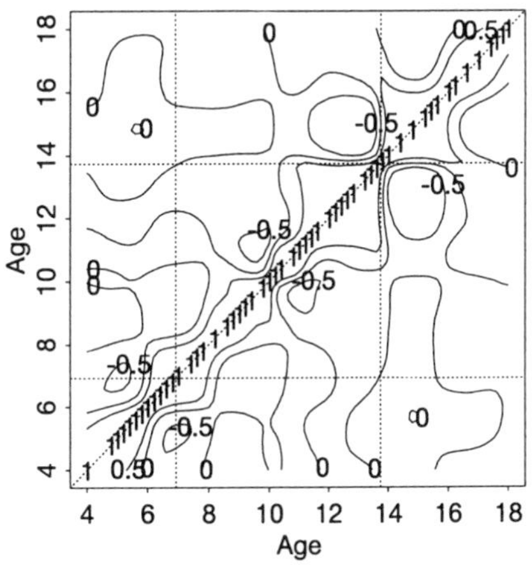

FIGURE 1.12. A contour plot of the correlation surface for the acceleration in height of the boys in the Fels Growth Study after registration. The centre of the pubertal growth spurt is at 14 years, and that of the mid-spurt is at 6.5 years. These ages are indicated by the horizontal and vertical dashed lines.

1.7.2 Plotting pairs of derivatives

Helpful clues to the processes giving rise to functional data can often be found in the *relationships* between derivatives. For example, two functions exhibiting simple derivative relationships are frequently found as strong influences in functional data: the exponential function, $f(t) = C_1 + C_2 e^{\alpha t}$, satisfies the differential equation

$$Df = -\alpha(f - C_1)$$

and the sinusoid $f(t) = C_1 + C_2 \sin[\omega(t - \tau)]$ with phase constant τ satisfies

$$D^2 f = -\omega^2(f - C_1).$$

Plotting the first or second derivative against the function value explores the possibility of demonstrating a linear relationship corresponding to one of these differential equations. Of course, it is usually not difficult to spot these types of functional variation by plotting the data themselves. However, plotting the higher derivative against the lower is often more informative, partly because it is easier to detect departures from linearity than from other functional forms, and partly because the differentiation may expose effects not easily seen in the original functions.

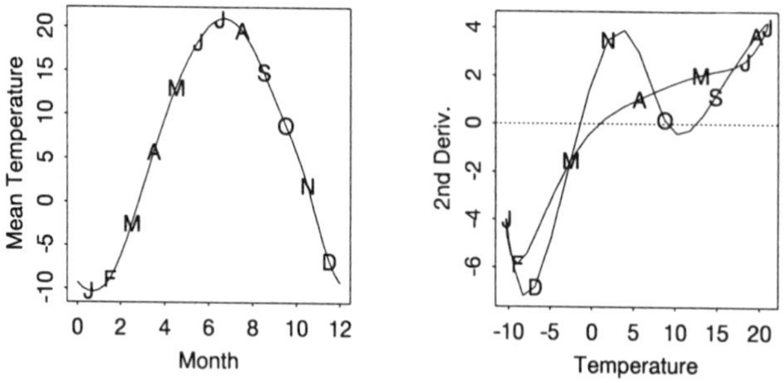

FIGURE 1.13. The left panel gives the annual variation in mean temperature at Montreal. The times of the mid-months are indicated by the first letters of the months. The right panel displays the relationship between the second derivative of temperature and temperature less its annual mean. Strictly sinusoidal or harmonic variation in temperature would imply a linear relationship.

Consider, for example, the variation in mean temperature Temp at Montreal displayed in the left panel of Figure 1.13. Casual inspection does indeed suggest a strongly sinusoidal relationship between temperature and month, but the right panel shows that things are not so simple. Although there is a broadly linear relationship between $-D^2$Temp and Temp after subtracting the mean annual temperature, there is obviously an additional systematic trend, which is more evident in the summer through winter months than in the spring. This plot greatly enhances the small departures from sinusoidal behaviour, and invites further attention.

1.7.3 Principal differential analysis

Chapter 14 takes up the question, novel in functional data analysis, of how to use derivative information in studying components of variation. An approach called *principal differential analysis* identifies important variance components by estimating a linear differential operator that will annihilate them. Linear differential operators, whether estimated from data or constructed from external modelling considerations, also play an important part in developing regularization methods more general than those in common use. Some of their aspects and advantages will be discussed in Chapter 15.

1.8 Concluding remarks

The last chapter of the book, Chapter 16, includes a discussion of some historical perspectives and bibliographic references not included in the main part of our development.

In the course of the book, we shall describe a considerable number of techniques and algorithms, to explain how the methodology we develop can actually be used in practice. We shall also illustrate our methodology on a variety of data sets drawn from various fields, including where appropriate the examples we have already introduced in this chapter. However, it is not our intention to provide a cook-book for functional data analysis, and still less a software manual.

In broad terms, we have a grander aim: to encourage readers to think about and understand functional data in a new way. The methods we set out are hardly the last word in approaching the particular problems, and we believe that readers will gain more benefit by using the principles we have laid down than by following our own suggestions to the letter. However, for those who would like access to the software we have used ourselves, a selection is available on a world-wide web site associated with the book. This web site will also be used to publicize related and future work by the authors and others, and to make available the data sets referred to in the book that we are permitted to release publicly. Finally, it will contain a list of errors in the text!

The home page for the book is accessible through the publisher's web site

```
http://www.springer-ny.com
```

or, at the time of writing, through either of the authors' home pages

```
http://www.psych.mcgill.ca/faculty/ramsay.html
http://www.statistics.bristol.ac.uk/~bernard
```

and will be updated from time to time.

2

Notation and techniques

2.1 Introduction

This chapter reviews topics that are notational and conceptual background to our main development of functional data analysis beginning in Chapter 3. However, many readers may prefer to skip straight to the next chapter and to refer back as necessary.

Further background review of useful techniques in linear and matrix algebra is provided in the Appendix. Among these are matrix decompositions, projections, and the constrained maximization of quadratic forms. We make occasional use of these tools, which will in any case be familiar to many readers, in our discussion of some of the more technical and algorithmic aspects of functional data analysis.

A basic message of this book is the value of thinking of a functional observation as a single datum rather than as a large set of data on its own. Notation facilitating this way of thinking is set out in this chapter. After some simple notational conventions in the next few pages, we introduce inner product notation, $\langle x, y \rangle$. Readers already familiar with this powerful device may want to move along, but should nevertheless be warned that we shall indulge ourselves in a few idiosyncracies.

Next we briefly review the extension to the functional setting of some standard concepts in linear algebra covered in the Appendix.

We close the chapter with a discussion of the multivariate linear model, to provide a useful link with some of the functional data analysis techniques discussed later on, especially for readers familiar with multivariate analysis. Particular attention is paid to design matrices not of full column rank, pointing forward to the concept of *regularization* discussed in detail in Chapter 4 and used repeatedly in later chapters of the book.

2.2 General notation

2.2.1 Scalars, vectors, functions and matrices

The reader should be warned that we try to use notation that brings out the basic structure of what is being done, and that this may entail the use of conventions a little unfamiliar at first sight. For example, we do not usually bother to distinguish in our notation between scalar quantities (numbers), vectors and functions. This means that a single symbol x can refer to a scalar, a vector with elements x_i, or a function with values $x(t)$.

If x is a vector or function, its elements or values x_i or $x(t)$ are usually scalars, but sometimes it is appropriate for the individual x_i or $x(t)$ to be a vector or function itself. The nervous reader should be assured that this convention is used only to clarify, rather than confuse, the discussion! In general, the context should always make clear when a symbol refers to a scalar, vector or function.

We follow more standard convention, however, in using bold capitals to refer to matrices, as in **X**.

It is often clearer to use longer strings of letters in a distinctive font to denote quantities more evocatively than standard notation allows. For example, we use names such as

- `Temp` for a temperature record,
- `Knee` for a knee angle,
- `LMSSE` for a squared-error fitting criterion for a linear model, and
- `RSQ` for a squared correlation measure.

We always use the notation x' for the transpose of a vector x. Note that $'$ is not used to denote differentiation. Instead, our notation for the derivative of order m of a function x is $D^m x$; this produces cleaner formulas than $d^m x/dt^m$. It stresses that differentiation is an *operator* that acts on a function x to produce another function Dx. We also use operators that act on functions in other ways, and it is convenient to use a consistent notation.

2.2.2 Combining quantities

We want to define familiar descriptive statistics, such as the mean vector and the variance-covariance matrix, in a way that works equally well for multivariate and functional data. This emphasizes the connections and similarities between multivariate and functional data analysis in a way that helps the reader to draw on experience with multivariate data. A pivotal role is played by *inner product* notation, which we discuss in Section 2.3, after briefly considering *addition.*

An advance in mathematical notation occurs when we separate the name for an operation from explicit instructions on how to carry it out. Consider, for example, the operation +. Suppose one opens a mathematics book at a random page and discovers the expression $x + y$. One might imagine that everyone would always mean the same by $x + y$, but a moment's thought shows that computing the sum can involve very different techniques depending on whether x and y are real numbers, complex numbers, vectors, matrices of the same dimensions or functions. What really counts is that we can assume that any author who uses the symbol + means an operation obeying the basic properties of addition, such as $x + y = y + x$, $(x + y) + z = x + (y + z)$, $(x + y)z = xy + xz$ if multiplication is defined, and so on. The author assumes that we ourselves can actually carry out the operation involved, or in some exotic situations he furnishes us with detailed instructions. The notation $x + y$ allows the basic structure of addition to be assumed, almost subconsciously, leaving the details to be supplied in any particular case, if necessary.

2.3 Inner products $\langle x, y \rangle$

2.3.1 Some specific examples

We now discuss a generic notation for inner products, extending the familiar idea of the inner product of two vectors x and y. Consider the *Euclidean inner product* operation $x'y$, where x and y are vectors of the same length. The operation has the following simple properties:

Symmetry: $x'y = y'x$ for all x and y,

Positivity: $x'x \geq 0$ for all x, with $x'x = 0$ if and only if $x = 0$, and

Bilinearity: for all real numbers a and b, $(ax + by)'z = ax'z + by'z$ for all vectors x, y and z.

Of course, these properties follow from the instructions implied in the definition

$$x'y = \sum_i x_i y_i. \tag{2.1}$$

But it is important to note that the Euclidean inner product operation is of critical importance in multivariate data analysis *because* of the properties of symmetry, positivity and bilinearity, which can therefore be considered fundamentally more significant than the definition (2.1) itself.

This basic role of symmetry, positivity and bilinearity is further emphasized when we realize that $x'\mathbf{W}y$, where $\mathbf{W}$ is a positive definite matrix of appropriate order, also has these properties and, indeed, can be used almost anywhere that we use $x'y$. So, for example, we use $x'\Sigma^{-1}y$, where Σ is a population covariance matrix, to define the multivariate normal distribution, to compute Mahalanobis distances, to define generalized least squares estimates instead of ordinary least squares, and many other useful things.

Now suppose that x and y are not vectors but rather functions with values $x(t)$. The natural functional counterpart to $x'y$ is $\int x(t)y(t)\,dt$, replacing the sum in (2.1) by an integral. Again, we have an operation on two functions x and y that is denoted by presenting the instructions for computing its value, but we know that this, too, is symmetric in x and y, linear in either function, and satisfies the positivity requirement. The same conclusions can be drawn for the operation $\int \omega(t)x(t)y(t)\,dt$, where ω is a strictly positive weight function, and indeed for the more general operation $\int\int \omega(s,t)x(s)y(t)\,ds\,dt$ if ω is strictly positive-definite, which simply means that the positivity requirement for the inner product is satisfied.

It should be clear by now that we can achieve a great leap forward in generality by using a common notation for these various real-valued operations that is understood to imply symmetry, positivity and bilinearity, without bothering with the details of the computation. We call such an operation an *inner product*, and we use the generic notation $\langle x, y\rangle$ for the inner product of x and y. The fundamental properties of an inner product are

Symmetry: $\langle x, y\rangle = \langle y, x\rangle$ for all x and y,

Positivity: $\langle x, x\rangle \geq 0$ for all x, with $\langle x, x\rangle = 0$ if and only if $x = 0$, and

Bilinearity: for all real numbers a and b, $\langle ax + by, z\rangle = a\langle x, z\rangle + b\langle y, z\rangle$ for all vectors x, y and z.

Note that bilinearity in the second argument follows from symmetry and bilinearity in the first.

2.3.2 General properties: association, size, angle, distance

We can think of the inner product as defining a scalar measure of *association* between pairs of quantities x and y. The symmetric nature of the measure means that , as we would usually require, it is invariant with respect to the order of the quantities. Bilinearity means that changing the scale of either argument has the same scale effect on the measure of association, and that the measure of association of one quantity with the sum of two others is the sum of the individual measures of association; these are both natural properties for a measure of association to have.

Positivity means that the inner product of any x with itself is essentially a measure of its *size*. The positive square root of this size measure is called the *norm* of x, written $\|x\|$, so that

$$\|x\|^2 = \langle x, x \rangle \tag{2.2}$$

with $\|x\| \geq 0$. In the special case where x is an n-vector and the inner product is the Euclidean inner product (2.1), the norm of x is simply the length of the vector measured in n-dimensional space. In the case of a function f, a basic type of norm is $\|f\| = \sqrt{\int f^2}$, called its $\mathcal{L}^2$ norm.

Whatever inner product is used, the standard properties of inner products lead to the following properties of the norm:

1. $\|x\| \geq 0$ and $\|x\| = 0$ if and only if $x = 0$.
2. $\|ax\| = |a|\,\|x\|$ for all real numbers a.
3. $\|x + y\| \leq \|x\| + \|y\|$.

From the properties of the inner product also follows the *Cauchy-Schwarz inequality*,

$$|\langle x, y \rangle| \leq \|x\|\,\|y\| = \sqrt{\langle x, x \rangle \langle y, y \rangle}.$$

This inequality links the inner product with the derived size measure or norm, and also leads to the *cosine inequality*,

$$-1 \leq \langle x, y \rangle / (\|x\|\,\|y\|) \leq 1.$$

The cosine inequality links the inner product to the geometrical concept of *angle*; the angle between x and y can be defined as the angle θ such that

$$\cos\theta = \frac{\langle x, y\rangle}{\|x\|\|y\|}.$$

Where x and y are n-vectors and the inner product is Euclidean inner product, θ is the angle between x and y in the usual geometric sense. Similarly, the cosine of the angle between two functions f and g can be defined as $\int fg/\sqrt{(\int f^2)(\int g^2)}$. The use of the cosine inequality to justify the idea of the angle between two vectors or functions further illuminates the notion that $\langle x, y\rangle$ is an association measure. Once we have obtained a scale-invariant coefficient by dividing by $\|x\|\|y\|$, we have a useful index of the extent to which x and y are measuring the same thing.

The particular relation $\langle x, y\rangle = 0$, called *orthogonality*, implies that x and y can be considered as being at right angles to one another. Because of bilinearity, orthogonality remains unchanged under any rescaling of either quantity. Orthogonality plays a key role in the operation of *projection* that is discussed in Section 2.4.1.

From the inner product, we also derive a measure of *distance* between x and y

$$d_{xy} = \|x - y\| = \sqrt{\langle x - y, x - y\rangle}$$

that has extremely wide applications; again, in the Euclidean case, distance corresponds to the usual geometric definition.

Thus, the simple algebraic properties of symmetry, positivity and bilinearity of the inner product lead easily to very useful definitions of the size of a quantity x, and of the angle and distance between x and y. We can be confident that, no matter how we define $\langle x, y\rangle$ in a particular application, the essential characteristics of these three measures remain unchanged.

The nature of the inner product depends on something more fundamental about x and y: They are elements of a *vector space* in which elements can be added or multiplied by real numbers to yield new vectors, and in which addition distributes with respect to scalar multiplication. The ensemble of a vector space and an associated inner product is called an *inner product space*.

Finally, of the three properties, only symmetry and bilinearity are really crucial. We can often get by with relaxing positivity to the weaker condition that $\langle x, x\rangle \geq 0$, so that $\langle x, x\rangle$ may be zero for some x's that are not themselves zero. Then the inner product is called a *semi-inner product* and the norm a *seminorm*. Most properties of inner products remain true for semi-inner products.

2.3.3 *Descriptive statistics in inner product notation*

As an example of how inner products can work for us, we consider how standard descriptive statistics can be expressed in inner product notation. Consider the space of possible univariate samples $x = (x_1, \ldots, x_N)$ of size N. Define the inner product to be the Euclidean inner product

$$\langle x, y \rangle = \sum_i x_i y_i = x' y.$$

Let 1 indicate the vector of size N all of whose elements are unity. Then some familiar univariate descriptive statistics become

Mean: $\bar{x} = N^{-1} \langle x, 1 \rangle$. Note that $\bar{x}$, being a multiple of an inner product, is a scalar and not a vector. The vector of length N all of whose elements are $\bar{x}$ is $\bar{x}1$.

Variance: $s_x^2 = N^{-1} \langle x - \bar{x}1, x - \bar{x}1 \rangle = N^{-1} \|x - \bar{x}1\|^2$.

Covariance: $s_{xy} = N^{-1} \langle x - \bar{x}1, y - \bar{y}1 \rangle$.

Correlation: $r_{xy} = s_{xy} / (s_x s_y)$.

It is easy to show that the covariance s_{xy} is itself a semi-inner product between x and y. Then it is an immediate consequence of the cosine inequality that the correlation coefficient satisfies the well-known *correlation inequality*

$$-1 \le r_{xy} \le 1.$$

Now suppose we stop using the Euclidean inner product but instead go for

$$\langle x, y \rangle = \sum_i w_i x_i y_i,$$

where w_i is a nonnegative weight to be applied to observation i. What difference would this make? None at all, except of course we must now divide by the constant $\sum_i w_i$ instead of N in defining $\bar{x}, s_x^2$, and s_{xy}. The essential characteristics of these statistics depend on the characteristics of the inner product, not on precisely how it is defined. Of course, the weighting affects the values of the statistics, but the essential meanings of the various descriptive statistics, for example as measures of location, scale and dependence, remain basically unchanged.

We can generalize this idea further: Suppose that the sequence of observations is known to be correlated, with covariance matrix Σ. Then we can use $\langle x, y \rangle = x' \Sigma^{-1} y$ to provide a basis for descriptive statistics that compensate for the known covariance structure on the observations.

Now consider these same statistics in the context of x as a function with values $x(t)$, where argument t takes values within some real interval such as $[0, T]$. Thus the index i taking N possible values has been replaced by the index t taking an infinity of values. Define the inner product as

$$\langle x, y \rangle = \int_0^T x(t)y(t)\,dt,$$

where we assume that the functions are sufficiently well behaved that the integral is always defined and finite. Then the various descriptive statistics continue to be defined as above, except that we divide by $\int_0^T dt = T$ instead of N and the vector $\mathbf{1}$ is replaced by the function $\mathbf{1} = 1(t)$ which takes the value of unity for all t. In the functional case, $\bar{x}$ becomes the mean level of the function x, s_x^2 becomes a measure of its variation about its mean level, and s_{xy} and r_{xy} measure the correspondence between the variation of x and y. Moving to

$$\langle x, y \rangle = \int_0^T \omega(t)x(t)y(t)\,dt$$

for some positive weight function ω and dividing by $\int \omega(t)\,dt$ really would not change these interpretations in any essential way, except that different parts of the range of t would be regarded as of different importance.

Finally, we note that even the divisors in these statistics can be defined in inner product terms, meaning that our fundamental descriptive statistics can be written in the unifying form

$$\begin{aligned}
\bar{x} &= \langle x, \mathbf{1} \rangle / \|\mathbf{1}\|^2, \\
s_x^2 &= \|x - \bar{x}\mathbf{1}\|^2 / \|\mathbf{1}\|^2, \text{ and} \\
s_{xy} &= \langle x - \bar{x}\mathbf{1}, y - \bar{y}\mathbf{1} \rangle / \|\mathbf{1}\|^2.
\end{aligned}$$

2.3.4 *Some extended uses of inner product notation*

In this book, we take the somewhat unorthodox step of using inner product notation to refer to certain *linear operations* that, strictly speaking, do not fall within the rubric of inner products.

So far in our discussion, the result of an inner product has always been a single real number. One way in which we extend our notation is the following. Let $x = (x_1, \ldots, x_m)'$ be a vector of length m, each element of which is an element of some vector space, whether finite-dimensional or functional. Then the notation $\langle x, y \rangle$, where y is a single element of the same space, indicates the m-vector whose elements are $\langle x_1, y\rangle, \ldots, \langle x_m, y\rangle$. Furthermore, if y is similarly a vector of length, say, n, then the notation $\langle x, y'\rangle$ defines the *matrix* with m rows and n columns containing the values $\langle x_i, y_j\rangle$, $i = 1, \ldots, m$, $j = 1, \ldots, n$. We use this convention only in situations where the context should make clear whether x and/or y are vectors of elements of the space in question.

In the functional context, we sometimes write

$$\langle z, \beta\rangle = \int z(s)\beta(s)\, ds$$

even when the functions z and β are not in the same space. We hope that the context of this use of inner product notation will make clear that a true inner product is not involved in this case. The alternative would have been the use of different notation such as (z, β), but we considered that the possibilities of confusion justified avoiding this convention.

An important property is that $\langle z, \beta\rangle$ is always a *linear operator* when regarded as a function of either of its arguments; generally speaking, a linear operator on a function space is a mapping A such that, for all f_1 and f_2 in the space and for all scalars a_1 and a_2, $A(a_1 f_1 + a_2 f_2) = a_1 A f_1 + a_2 A f_2$.

2.4 Further aspects of inner product spaces

We briefly review two further aspects of inner product spaces that are useful in our later development. The first of these is the concept of *projection*. This generalizes ideas about projection matrices which are reviewed in detail in the Appendix. In the present discussion we do not go into any great technical detail.

2.4.1 Projections

Let $u_1, \ldots, u_n$ be any n elements of our space, and let $\mathcal{U}$ be the subspace consisting of all possible linear combinations of the u_i. We can characterize the subspace $\mathcal{U}$ by using suitable vector notation. Let u be the n-vector whose elements are the $u_1, \ldots, u_n$. Then every member of $\mathcal{U}$ is of the form $u'c$ for some real n-vector c.

Associated with the subspace $\mathcal{U}$ is the *orthogonal projection onto* $\mathcal{U}$, defined to be a linear operator P with the following properties:

1. For all z, the element Pz falls in $\mathcal{U}$, and so is a linear combination of the functions $u_1, \ldots, u_n$.

2. If y is in $\mathcal{U}$ already, then $Py = y$.

3. For all z, the *residual* $z - Pz$ is orthogonal to all elements v of $\mathcal{U}$.

From the first two of these properties, it follows at once that $PP = P^2 = P$. From the third property, it is easy to show that the operator P maps each element z to its *nearest* point in $\mathcal{U}$, distance being measured in terms of the norm. This makes projections very important in statistical contexts such as least squares estimation. We give a justification of this property of orthogonal projections in Section A.2.3 of the Appendix, where we also address the question of how to carry out the projection P in practice.

2.4.2 *Quadratic optimization*

Some of our functional data analysis methodology require the solution of a particular kind of constrained optimization problem. Suppose that A is a linear operator on a function space satisfying the condition

$$\langle x, Ay \rangle = \langle Ax, y \rangle \text{ for all } x \text{ and } y.$$

Such an operator is called a *self-adjoint* operator.

Now consider the problem of maximizing $\langle x, Ax \rangle$ subject to the constraint $\|x\| = 1$. In Section A.3 of the Appendix, we set out results relating this optimization problem to the eigenfunction/eigenvalue problem $Au = \lambda u$. We go on to consider the more general problem of maximizing $\langle x, Ax \rangle$ subject to a constraint on $\langle x, Bx \rangle$ for a second self-adjoint operator B. We do not go into further detail here but refer the reader to the Appendix.

2.5 The multivariate linear model

We now return to a more statistical topic. A review of the multivariate linear model may be helpful, both to fix ideas and notation, and because some of the essential concepts transfer to functional contexts without much more than a change of notation. But a slight change of perspective is helpful on what the design matrix means. Moreover, a notion used repeatedly for functional data is *regularization*, which we introduce in Section 2.6 within the multivariate context.

2.5.1 *Linear models from a transformation perspective*

Let $\mathbf{Y}$ be a $N \times p$ matrix of dependent variable observations, $\mathbf{Z}$ be a $N \times q$ matrix, and $\mathbf{B}$ be a $q \times p$ matrix. In classical terminology, $\mathbf{Z}$ is the *design matrix* and $\mathbf{B}$ is a matrix of *parameters.*

The multivariate linear model is

$$\mathbf{Y} = \mathbf{ZB} + \mathbf{E}. \tag{2.3}$$

At least at the population level, the rows of the disturbance or residual matrix $\mathbf{E}$ are often thought of as independent samples from a common population of p-variate observations with mean 0 and finite covariance matrix $\mathbf{\Sigma}$.

Although in many contexts it is appropriate to think of the columns of $\mathbf{Z}$ as corresponding to variables, it is better for our purposes to take the more general view that $\mathbf{Z}$ represents a linear transformation that maps matrices $\mathbf{B}$ into matrices with the dimensions of $\mathbf{Y}$. This can be indicated by the notation

$$\mathbf{Z} : R^{q \times p} \to R^{N \times p}$$

The space of all possible transformed values $\mathbf{ZB}$ then defines a subspace of $R^{N \times p}$, denoted by $R(\mathbf{Z})$ and called the *range space* of $\mathbf{Z}$.

2.5.2 *The least squares solution*

When it is assumed that the rows of the disturbance matrix $\mathbf{E}$ are independent, each with covariance matrix $\mathbf{\Sigma}$, the natural inner product to use in the observation space $R^{N \times p}$ is

$$\langle \mathbf{X}, \mathbf{Y} \rangle = \text{trace}\, \mathbf{X\Sigma}^{-1}\mathbf{Y}' = \text{trace}\, \mathbf{Y}'\mathbf{X\Sigma}^{-1} \tag{2.4}$$

for $\mathbf{X}$ and $\mathbf{Y}$ in $R^{N \times p}$. Then we measure the goodness of fit of any parameter matrix $\mathbf{B}$ to the observed data $\mathbf{Y}$ by the corresponding squared norm

$$\texttt{LMSSE}(\mathbf{B}) = \|\mathbf{Y} - \mathbf{ZB}\|^2 = \text{trace}\, (\mathbf{Y} - \mathbf{ZB})'(\mathbf{Y} - \mathbf{ZB})\mathbf{\Sigma}^{-1}. \tag{2.5}$$

For the moment, suppose that the matrix $\mathbf{Z}$ is of full column rank, or that $N \geq q$ and the columns of $\mathbf{Z}$ are independent. A central result on the multivariate linear model is that the matrix $\hat{\mathbf{B}}$ that minimizes $\texttt{LMSSE}(\mathbf{B})$ is given by

$$\hat{\mathbf{B}} = (\mathbf{Z}'\mathbf{Z})^{-1}\mathbf{Z}'\mathbf{Y}. \tag{2.6}$$

The corresponding predictor of $\mathbf{Y}$ is given by

$$\hat{\mathbf{Y}} = \mathbf{Z}\hat{\mathbf{B}} = \mathbf{Z}(\mathbf{Z}'\mathbf{Z})^{-1}\mathbf{Z}'\mathbf{Y}. \tag{2.7}$$

The matrix $\hat{\mathbf{Y}}$ can be thought of as the matrix in the subspace $R(\mathbf{Z})$ that minimizes $\|\mathbf{Y} - \hat{\mathbf{Y}}\|^2$ over all possible approximations $\hat{\mathbf{Y}} = \mathbf{Z}\mathbf{B}$ falling in $R(\mathbf{Z})$.

Note that the least squares estimator $\hat{\mathbf{B}}$ and the best linear predictor $\hat{\mathbf{Y}}$ do not depend on the variance matrix Σ, even though the fitting criterion $\mathsf{LMSSE}(\mathbf{B})$ does. It turns out that when the details of the minimization of $\mathsf{LMSSE}(\mathbf{B})$ are carried through, the variance matrix Σ cancels out. But if there are covariances among errors or residuals *across* observations, contained in the variance-covariance matrix Γ, say, then the inner product (2.4) becomes

$$\langle \mathbf{X}, \mathbf{Y} \rangle = \text{trace } \mathbf{Y}'\Gamma^{-1}\mathbf{X}\Sigma^{-1}.$$

Using this inner product in the definition of goodness of fit, the estimator of $\mathbf{B}$ and the best predictor of $\mathbf{Y}$ become

$$\hat{\mathbf{B}} = (\mathbf{Z}'\Gamma^{-1}\mathbf{Z})^{-1}\mathbf{Z}'\Gamma^{-1}\mathbf{Y} \quad \text{and} \quad \hat{\mathbf{Y}} = \mathbf{Z}(\mathbf{Z}'\Gamma^{-1}\mathbf{Z})^{-1}\mathbf{Z}'\Gamma^{-1}\mathbf{Y}.$$

Thus, the optimal solution does depend on how one treats errors *across* observations.

2.6 Regularizing the multivariate linear model

One of the major themes of this book is *regularization*, and for readers familiar with multivariate analysis, it may be helpful to introduce this idea in the multivariate context first. Others, especially those who are familiar with curve estimation already, may prefer to omit this section.

Now suppose that we are dealing with an underdetermined problem, where $q > N$ and the matrix $\mathbf{Z}$ is of full row rank N. This means that the range space $R(\mathbf{Z})$ is the whole of $R^{N \times p}$.

2.6.1 *Definition of regularization*

Regularization involves attaching a *penalty term* to the basic squared-error fitting criterion:

$$\mathsf{LMSSE}_\lambda(\mathbf{B}) = \|\mathbf{Y} - \mathbf{Z}\mathbf{B}\|^2 + \lambda \times \mathsf{PEN}(\mathbf{B}). \tag{2.8}$$

The purpose of the penalty term $\mathsf{PEN}(\mathbf{B})$ is to require that the estimated value of $\mathbf{B}$ not only yields a good fit in the sense of small $\|\mathbf{Y} - \mathbf{Z}\mathbf{B}\|^2$, but also that some aspect of $\mathbf{B}$ captured in the function PEN is kept under control. The positive penalty parameter λ quantifies the relative importance of these two aims. If λ is large, then we are particularly concerned with keeping $\mathsf{PEN}(\mathbf{B})$ small, and getting a good fit to the data is only of secondary importance. If λ is small, then we are not so concerned about the value of $\mathsf{PEN}(\mathbf{B})$.

One example of this type of regularization is the *ridge regression* technique, often used to stabilize regression coefficient estimates in the presence of highly collinear independent variables. In this case, what is penalized is the size of the regression coefficients themselves, in the sense that $\mathsf{PEN}(\mathbf{B}) = \text{trace}(\mathbf{B}'\mathbf{B})$, the sum of squares of the entries of $\mathbf{B}$. The solution to the minimization of $\mathsf{LMSSE}_\lambda(\mathbf{B})$ is then

$$\mathbf{B} = (\mathbf{Z}'\mathbf{Z} + \lambda\mathbf{I})^{-1}\mathbf{Z}'\mathbf{Y}.$$

As λ approaches zero, $\mathbf{B}$ approaches the least squares solution described in Section 2.5, but as λ grows, $\mathbf{B}$ approaches zero. Thus, ridge regression is said to shrink the solution towards zero.

2.6.2 Hard-edged constraints

One way to obtain a well-determined problem is to place constraints on the matrix $\mathbf{B}$. For example, consider the model where it is assumed that the coefficients in each column of $\mathbf{B}$ form a constant vector, so all we have to do is to estimate a single number for each column. If we define the $(q-1) \times q$ matrix $\mathbf{L}$ to have $L_{ii} = 1$ and $L_{i,i+1} = -1$ for each i, and all other entries zero, then our assumption about $\mathbf{B}$ can be written as the constraint

$$\mathbf{LB} = 0. \tag{2.9}$$

For the elements of $\mathbf{B}$ to be identifiable from the observed data, the design matrix $\mathbf{Z}$ has to satisfy the condition

$$\mathbf{Z1} \neq 0, \tag{2.10}$$

where $\mathbf{1}$ is a vector of q unities.

The transformation $\mathbf{L}$ reduces multiples of the vector $\mathbf{1}$ exactly to zero. The identifiability condition (2.10) can be replaced by the condition that the zero vector is the only q-vector b such that both $\mathbf{L}b$ and $\mathbf{Z}b$ are zero. Equivalently, the matrix $[\mathbf{Z}' \;\; \mathbf{L}']$ is nonsingular.

2.6.3 Soft-edged constraints

Instead of enforcing the hard-edged constraint $\mathbf{LB} = 0$, we may wish to let the coefficients in any column of $\mathbf{B}$ vary, but not more than really necessary, by exploring compromises between the rank-one extreme implied by (2.9) and a completely unconstrained underdetermined fit. We might consider this a soft-edged constraint, and it can be implemented by a suitable regularization procedure. If we define

$$\mathsf{PEN}(\mathbf{B}) = \|\mathbf{LB}\|^2 = \text{trace}(\mathbf{B}'\mathbf{L}'\mathbf{LB}), \tag{2.11}$$

then the penalty $\mathsf{PEN}(\mathbf{B})$ quantifies how far the matrix $\mathbf{B}$ is from satisfying the constraint $\mathbf{LB} = 0$.

The regularized estimate of $\mathbf{B}$, obtained by minimizing the criterion (2.8), now satisfies

$$(\mathbf{Z}'\mathbf{Z} + \lambda\mathbf{L}'\mathbf{L})\mathbf{B} = \mathbf{Z}'\mathbf{Y}. \tag{2.12}$$

For any $\lambda > 0$, a unique solution for $\mathbf{B}$ requires the nonsingularity of the matrix $[\mathbf{Z}' \quad \mathbf{L}']$, precisely the condition for identifiability of the model subject to the constraint (2.9).

In the limit as the parameter $\lambda \to \infty$, the penalized fitting criterion (2.8) automatically enforces on $\mathbf{B}$ the one-dimensional structure $\mathbf{LB} = 0$. On the other hand, in the limit $\lambda \to 0$, no penalty at all is applied, and $\mathbf{B}$ takes on whatever value results in minimizing the error sum of squares to zero, because of the underdetermined character of the problem. Thus, from the regularization perspective, the constrained estimation problem $\mathbf{LB} = 0$ that arises frequently in linear modelling designs is simply an extreme case of the regularization process where $\lambda \to \infty$.

We have concentrated on a one-dimensional constrained model, corresponding to a $(q - 1) \times q$ matrix $\mathbf{L}$, but of course the ideas can be immediately extended to nonsingular $s \times q$ constraint matrices $\mathbf{L}$ that map a q-vector into a space of vectors of dimension $s \leq q$. In this case, the constrained model is of dimension $q - s$. Note also that the specification of the matrix $\mathbf{L}$ corresponding to any particular constrained model is not unique. If $\mathbf{L}$ is specified differently the regularized estimates are in general different.

Finally, we note in passing that Bayesian approaches to regression, in which a multivariate normal prior distribution is proposed for $\mathbf{B}$, can also be expressed in terms of a penalized least squares problem of the form (2.8). For further details see, for example, Kimeldorf and Wahba (1970), Wahba (1978) or Silverman (1985).

3

Representing functional data as smooth functions

3.1 Introduction

3.1.1 Observed functional data

The basic philosophy of functional data analysis is that we should think of observed data functions as single entities, rather than merely a sequence of individual observations. The term functional refers to the intrinsic structure of the data rather than to their explicit form. But in practice, functional data are usually observed and recorded discretely. A record of a functional observation x consists of n pairs (t_j, y_j), where y_j is a recording or observation of $x(t_j)$, a snapshot of the function at argument value t_j.

In this chapter, we consider some techniques for converting raw functional data into true functional form. Since some observational noise is part of most data, the functional representation of raw data usually involves some smoothing, and so we review various smoothing techniques. Special attention is given to estimating derivatives, since these are important in many functional data analyses. In accordance with the notational conventions we have already established, let $D^m x$ indicate the mth derivative of a univariate function x. The value of the mth derivative for argument value t is indicated by $D^m x(t)$.

What would it mean for a functional observation x to be known in true functional form? We do not mean that x is actually recorded for every value of t, because that would involve storing an uncountable number of values! Rather, we mean that we posit the existence of a function x, based on the given data, implying that, in principle, we can evaluate x at any point t, and in addition that we can evaluate any of its derivatives $D^m x$ that exist at t. But because only discrete values are actually available, evaluating $x(t)$ and $D^m(t)$ at any arbitrary value T will involve some form of interpolation or smoothing of these discrete values.

In general, we are concerned with a collection of functional data, rather than just a single function x. Specifically, the record or observation of the function x_i might consist of n_i pairs (t_{ij}, y_{ij}), $j = 1, \ldots, n_i$. It may be that the argument values t_{ij} are the same value for each record, but they may also vary from record to record. We conceptualize the problem as functional rather than multivariate because it is assumed that functions x_i lie behind the data. In general, we consider the reconstruction of the functional observations x_i one by one. Therefore, in this chapter, we simplify notation and assume that a single function x is being observed.

It is always assumed that the range of values of interest for the argument t is a bounded interval $\mathcal{T}$, and, implicitly or explicitly, that x satisfies reasonable continuity or smoothness conditions on $\mathcal{T}$. Without some conditions of this kind, it is impossible to draw any inferences at all about values $x(t)$ for any points t apart from actual observation points. Sometimes the argument is cyclic, for instance when t is the time of year, and this means that the functions satisfy *periodic boundary conditions*, where the function x at the beginning of the interval $\mathcal{T}$ picks up smoothly from the values of x at the end.

3.1.2 Sampling and observational error

Smoothness, in the sense possessing of a certain number of derivatives, is a property of the latent function x, and may not be at all obvious in the raw data vector $y = (y_1, \ldots, y_n)$ because of observational error or noise that may be imposed on the underlying signal by aspects of the measurement process. In modelling terms, we write

$$y_j = x(t_j) + \epsilon_j \tag{3.1}$$

where the disturbance, error, perturbation or otherwise exogenous term ϵ_j contributes a roughness to the raw data. One of the tasks in representing the raw data as functions may be to attempt to filter out this noise as efficiently as possible. In some cases we may pursue an alternative strategy of leaving the noise in the data, and, instead, require smoothness of the results of our analysis, rather than of the data that are analysed.

The standard statistical model for the ϵ_j is that they are independently distributed with zero mean and finite variance. It is also convenient to assume that a constant variance σ^2 is common to these distributions. But in special cases, we must take explicit account of variance nonhomogeneity and covariances of the ϵ_j's for neighbouring argument values. Finally, we should keep in mind the possibility that errors or disturbances multiply rather than add, in which case it will be more convenient to work with the logarithms of the data.

The sampling rate or resolution of the raw data is a key determinant of what is possible in functional data analysis. The most important aspect is an essentially local property of the data, the density of the argument values t_j relative to the amount of curvature in the data, rather than simply the number n of argument values. Curvature at argument t is usually measured by the size $|D^2x(t)|$ of the second derivative. Where curvature is high, it is essential to have enough points to estimate the function effectively. What is enough? This depends on the amount of error ϵ_j; when the error level is small and the curvature is mild, as is the case for the Canadian temperature data in Figure 1.3, we can get away with a low sampling rate. The human growth data in Figure 1.1 have moderately low error levels (about 0.3% of adult height), but the curvature in the second derivative functions is fairly severe, so that the sampling rate for these data is barely adequate for making inferences about growth acceleration. The gait data in Figure 1.5 exhibit little error, and thus the sampling rate of 20 values per cycle is sufficient for most purposes.

Figure 3.1 provides an interesting example of functional data. The letters "fda" were written on a flat surface by one of the authors. The pen positions were recorded by an Optotrak system that gives the position of an infrared-emitting diode in three-dimensional space 600 times per second. The X and Y position functions `ScriptX` and `ScriptY` are plotted separately in Figure 3.2, and we can see that the error level is too small to be visible. The total event took about 2.3 seconds, and the plotted functions each have 1401 discrete values. This may seem like a lot, but the curvature is rather high in places, and it turns out that, even with the small error level, this level of resolution is important.

We shall investigate various ways in which the discrete observations y_j can be represented by an appropriately smooth function x, paying particular attention to estimating its derivative values. But first, some words of caution about estimating derivatives directly from raw data. Because the observed function looks reasonably smooth, one might be

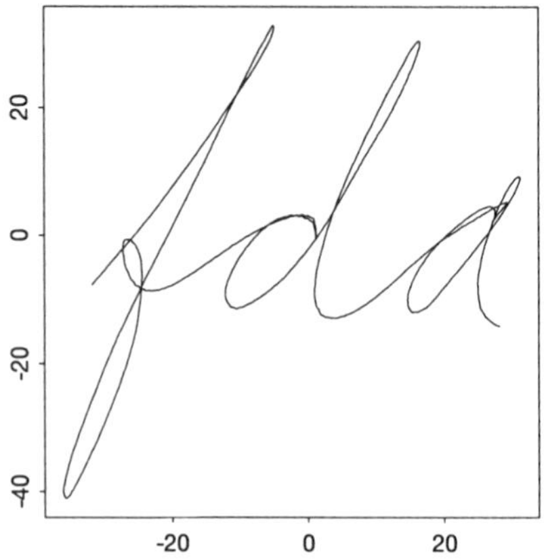

FIGURE 3.1. A sample of handwriting in which the X and Y coordinates are recorded 600 times per second.

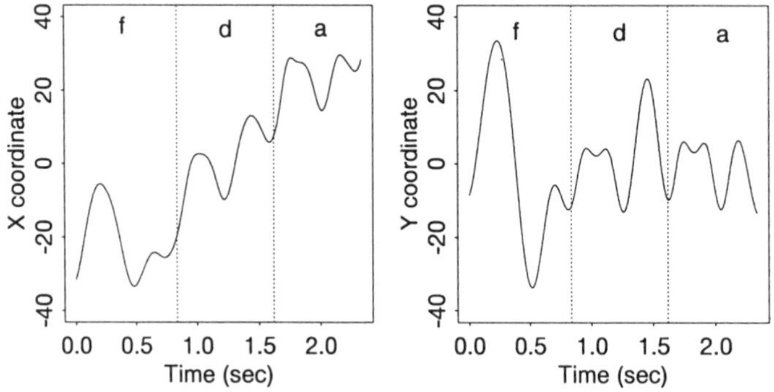

FIGURE 3.2. The X and Y coordinates for the handwriting sample plotted separately. Note the strongly periodic component with from two to three cycles per second.

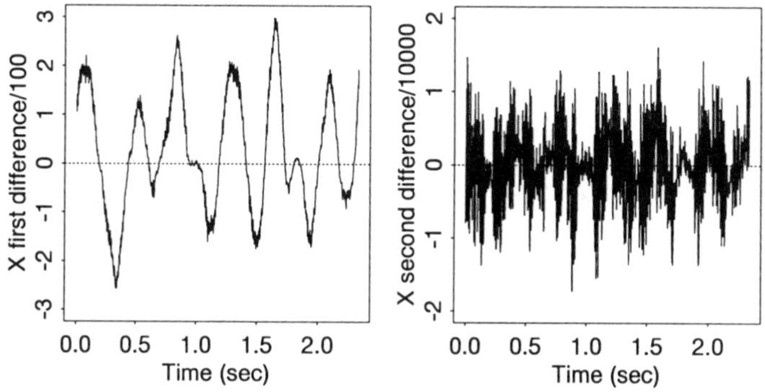

FIGURE 3.3. The first and second central differences for the X coordinate for the handwriting sample. The high sampling rate causes differencing to greatly magnify the influence of noise.

tempted to use the first forward difference $(y_{j+1} - y_j)/(t_{j+1} - t_j)$ or the central difference $(y_{j+1} - y_{j-1})/[2(t_{j+1} - t_{j-1})]$ to estimate $Dx(t_j)$, but Figure 3.3 shows that the resulting derivative estimate for `ScriptX` is rather noisy. The second difference estimate of D^2`ScriptX` assuming equally spaced arguments,

$$D^2x(t_j) \approx (y_{j+1} + y_{j-1} - 2y_j)/(\Delta t)^2,$$

is shown in Figure 3.3 to be a disaster, displaying far too little signal relative to the noise. The reason for this is the high sampling rate for the data; taking differences between extremely close values magnifies the influence of error enormously. Note also the cautions of Press et al. (1992) on using simple differencing to estimate derivatives even when functions are available analytically. Although a common practice, derivative estimation by differencing is, in fact, seldom a good idea!

As a final word on observational error, we point out that the classical model (3.1) could be replaced by a model

$$y_j = x(t_j) + \epsilon(t_j), \tag{3.2}$$

where a noise *function* ϵ is added to the smooth signal x and then evaluated. We might assume that ϵ has the intuitive characteristics of white noise: mean zero, constant variance, and covariance zero for distinct argument values.

We may have a substantive reason to prefer the discrete noise model (3.1) to the functional noise model (3.2). For example, the growth data clearly have discrete or observational noise. From experiments, we know that independent measurements of the same child at a fixed time have a standard error of around 1.5 mm. On the other hand, the temperature data are estimates of a mean function based on a sample of about 30 annual functions, and can be reasonably viewed as having no functional error, since the error in a temperature measurement is a very small fraction of the variation in actual temperature, whether taken over scales of minutes, days, or years. In any case, discrete data do not offer any way of choosing between these models.

Although the techniques in this chapter are designed to filter out observational error in the functions themselves, such error, whether discrete or functional, is not the only model possible for the variation from one functional observation to another. Noise can also affect one or more derivatives. For example, the variation in the handwriting data, where the observational error is, after all, a small fraction of the signal, can be caused by variation in the forces applied to the pen by muscle contractions. By Newton's Third Law these forces will affect the position of the pen only indirectly through their effect on its acceleration. Our plot of pen position in Figure 3.1 would look smooth even if the acceleration were noisy, because integrating acceleration twice to give position would inevitably smooth the noise but not reduce it to zero. These considerations are particularly relevant to the discussion in Chapters 13 to 15, where the use of linear differential operators to analyse functional data is explored in detail. But generally we need to keep in mind that some types of noise will appear smooth at the level of observation.

Now we turn to a discussion of various smoothing methods designed for direct observational error. Our goal is to give enough information to those new to the topic to launch a functional data analysis. For more complete treatments of this large and rapidly evolving field, many of them concentrating on particular aspects and methods, we refer the reader to sources such as Eubank (1988), Fan and Gijbels (1996), Green and Silverman (1994), Härdle (1990), Hastie and Tibshirani (1990), Simonoff (1996) and Wand and Jones (1995). One particular class of methods, based on roughness penalties, plays a particular rôle in our development of functional data analysis methods, and this is discussed separately in Chapter 4.

Most of the methods we consider are linear smoothers, and it is useful to discuss this general class of smoothers first of all.

3.1.3 Linear smoothing

A *linear smoother* estimates the function value $x(t)$ by a linear combination of the discrete observations

$$\hat{x}(t) = \sum_{j=1}^{n} S_j(t)y_j. \tag{3.3}$$

The behaviour of the smoother at t is determined by the weights $S_j(t)$. These weights do not depend only on the argument value t at which the function is to be estimated and on the particular within-record index j. In general, there is some kind of dependence on the whole pattern of observation points $(t_1, \ldots, t_n)$, but this is not expressed explicitly in the notation.

Linear smoothers can be represented in matrix form. Suppose we have in mind a sequence $s_1 < s_2 < \ldots < s_m$ of *evaluation values* in $\mathcal{T}$ at which the function x is to be estimated. Note that the evaluation values need not be the same as the observation values t_j. Write $\hat{x}$ for the m-vector of values $x(s_i)$ and y for the vector of observed data y_j. We can then write

$$\hat{x} = \mathbf{S}y \tag{3.4}$$

where $\mathbf{S}_{ij} = S_j(s_i)$.

Many widely used smoothers are linear. The linearity of a smoother is a desirable feature for various reasons: The linearity property

$$\mathbf{S}(ay + bz) = a\mathbf{S}y + b\mathbf{S}z$$

is important for working out various properties of the smooth representation, and the simplicity of the smoother implies relatively fast computation. On the other hand, some nonlinear smoothers may be more adaptive to different behaviour in different parts of the range of observation, and may be robust to outlying observations. Smoothing by the thresholded wavelet transform, discussed in Section 3.2.6, is an important example of a nonlinear smoothing method.

Speed of computation can be critical; a smoother useful for a few hundred evaluation or data points can be completely impractical for thousands. Smoothers that require of the order of n, abbreviated $O(n)$, operations to compute n smoothed values $\hat{x}(s_i)$ are virtually essential for large n, but linear smoothers that are less efficient computationally may be highly desirable in other ways. If $\mathbf{S}$ is band-structured, meaning that only a small number K of values on either side of its diagonal in any row are nonzero, then $O(n)$ computation is assured. However, it is not only for band-structured $\mathbf{S}$ that $O(n)$ computations are possible; see, for example, the discussion of spline smoothing in Green and Silverman (1994) and of L-spline smoothing in Chapter 15 of this book.

3.2 Basis function methods

3.2.1 Least squares fitting of basis expansions

One of the most familiar smoothing procedures involves representing the function by a linear combination of K known basis functions ϕ_k,

$$x(t) = \sum_{k=1}^{K} c_k \phi_k(t). \tag{3.5}$$

The degree to which the data y_j are smoothed, rather than exactly reproduced or interpolated, is determined by the number K of basis functions. Assuming that the $n \times K$ matrix $\Phi = \{\phi_k(t_j)\}$ of basis function values at the observation points is of full rank, an exact representation is generally possible when $K = n$, in the sense that we can choose the coefficients c_k to yield $x(t_j) = y_j$ for each j.

The simplest linear smoother defined by a basis function expansion is obtained if we determine the coefficients of the expansion c_k by minimizing the least squares criterion

$$\text{SMSSE}(y|c) = \sum_{j=1}^{n} [y_j - \sum_{k=1}^{K} c_k \phi_k(t_j)]^2, \tag{3.6}$$

or, in matrix terms,

$$\text{SMSSE}(y|c) = (y - \Phi c)'(y - \Phi c) = \|y - \Phi c\|^2,$$

where the K-vector c contains the coefficients c_k. This criterion is minimized by the solution $c = (\Phi'\Phi)^{-1}\Phi' y$.

Assuming that the evaluation points are identical to the observation points, the smoothing matrix $\mathbf{S}$ is then given by

$$\mathbf{S} = \Phi(\Phi'\Phi)^{-1}\Phi'. \tag{3.7}$$

The smoothing matrix $\mathbf{S}$ in this case is an orthogonal projection matrix since it is symmetric and satisfies the idempotency relation $\mathbf{S}^2 = \mathbf{S}$. Thus, a least squares basis function smoother is simply an orthogonal projection of a record onto the space spanned by the columns of the basis matrix Φ.

More generally, if the evaluation points s_i are not necessarily the same as the observation points, define the matrix $m \times n$ matrix $\tilde{\Phi}$ to have elements $\phi_j(s_i)$. Then the smoothing matrix is given by

$$\mathbf{S} = \tilde{\Phi}(\Phi'\Phi)^{-1}\Phi'.$$

We can extend the least squares criterion to the form

$$\text{SMSSE}(y|c) = (y - \Phi c)'\mathbf{W}(y - \Phi c) = \|y - \Phi c\|^2_{\mathbf{W}}, \qquad (3.8)$$

where $\mathbf{W}$ is a known symmetric positive-definite matrix that allows for unequal weighting of squares and products of residuals. This extension can be important if we know that the variances of the disturbances ϵ_j are not constant, or that the disturbances are not independently distributed in some known way. The estimates of the coefficients c are then given by $(\Phi'\mathbf{W}\Phi)^{-1}\Phi'\mathbf{Wy}$. In the case where the evaluation arguments and the data arguments are the same, the corresponding smoothing matrix is then

$$\mathbf{S} = \Phi(\Phi'\mathbf{W}\Phi)^{-1}\Phi'\mathbf{W},$$

and is still an orthogonal projection operator since $\mathbf{WS}$ is symmetric. The mapping defined by $\mathbf{S}$ is often said to be a *projection in the metric* $\mathbf{W}$.

A central question is how to choose the order of the expansion K. Since a least squares basis function expansion is essentially a multiple regression problem, the large literature on selecting a good subset of variables from a larger pool offers considerable guidance. Nevertheless, the fact that K can take only integer values implies that control over smoothing may be comparatively crude. The sections on localized least squares methods and smoothing splines in this chapter indicate how finer control is possible.

When n is large, efficient computation is critical. There are three essential tasks in computing the estimates for general evaluation points:

1. Compute inner products, of which there are K in $\Phi'y$ and $K(K+1)/2$ in $\Phi'\Phi$.
2. Solve the linear system $\Phi'\Phi c = \Phi'y$.
3. Compute the m inner products $\tilde{\Phi}c$, where the matrix $\tilde{\Phi}$ contains the basis functions evaluated at the evaluation arguments.

Efficient and stable least squares algorithms can perform these calculations or their equivalents in $O[(n+m)K^2]$ operations, and this is acceptable provided K is small and fixed relative to n and m. But for large K it is extremely helpful, for both computational economy and numerical stability, if the cross-product matrix $\Phi'\Phi$ has a band structure such that nonzero values appear only in a fixed and limited number of positions on either side of the diagonal. A basis orthogonal with respect to summation over j is the extreme case because the cross-product matrix is diagonal; the coefficients of the least squares fit are then found by multiplying the data vector y by an explicit $K \times n$ matrix, without needing to solve any system of linear equations.

In the worst-case scenario, without a band structure and with K and m of $O(n)$, the computation is $O(n^3)$, which is unacceptable for n in the many hundreds. Thus efficient computation is essential for the handwriting data ($n = 1401$) and for the daily weather data.

3.2.2 *Choosing a good basis*

A desirable characteristic of basis functions is that they have features matching those known to belong to the functions being estimated. Ideally, a basis should be chosen to achieve an excellent approximation using a comparatively small value of K; not only does this imply less computation, but the coefficients themselves can become interesting descriptors of the data from a substantive point of view. Consequently, certain classic off-the-rack bases may, in fact, be ill-advised in some applications; there is no such thing as a good universal basis.

The choice of basis is particularly important for the derivative estimate

$$D\hat{x}(t) = \sum_{k=1}^{K} c_k D\phi_k(t). \tag{3.9}$$

Bases that work well for function estimation may give rather poor derivative estimates. This is because an accurate representation of the observations may force $\hat{x}$ to have small but high-frequency oscillations with horrible consequences for its derivatives. Put more positively, one of the criteria for choosing a basis may be whether or not one or more of the derivatives of the approximation behave reasonably.

Chapter 15 touches on tailoring a basis to fit a particular problem. For now, we discuss some popular bases that are widely used in practice.

3.2.3 *Fourier series*

Perhaps the best known basis expansion is provided by the Fourier series:

$$\hat{x}(t) = c_0 + c_1 \sin \omega t + c_2 \cos \omega t + c_3 \sin 2\omega t + c_4 \cos 2\omega t + \dots \quad (3.10)$$

defined by the basis $\phi_0(t) = 1, \phi_{2r-1}(t) = \sin r\omega t$, and $\phi_{2r}(t) = \cos r\omega t$. This basis is periodic, and the parameter ω determines the period $2\pi/\omega$, which is equal to the length of the interval $\mathcal{T}$ on which we are working. If the values of t_j are equally spaced on $\mathcal{T}$, then the basis is orthogonal in the sense that the cross-product matrix $\Phi'\Phi$ is diagonal, and can be made equal to the identity by dividing the basis functions by suitable constants, $\sqrt{n}$ for $j = 0$ and $\sqrt{n/2}$ for all other j.

The Fast Fourier transform (FFT) makes it possible to find all the coefficients extremely efficiently; when n is a power of 2 and the arguments are equally spaced, then we can find both the coefficients c_k and all n smooth values at $x(t_j)$ in $O(n \log n)$ operations. This is one of the features that has made Fourier series the traditional basis of choice for long time series.

Derivative estimation is simple since

$$D \sin r\omega t = r\omega \cos r\omega t$$

and

$$D \cos r\omega t = -r\omega \sin r\omega t.$$

This implies that the Fourier expansion of Dx has coefficients

$$(0, -\omega c_2, \omega c_1, -2\omega c_4, 2\omega c_3, \dots)$$

and of $D^2 x$ has coefficients

$$(0, -\omega^2 c_1, -\omega^2 c_2, -4\omega^2 c_3, -4\omega^2 c_4, \dots).$$

Similarly, we can find the Fourier expansions of higher derivatives by multiplying individual coefficients by suitable powers of $r\omega$, with appropriate sign changes and interchanges of sine and cosine coefficients.

The Fourier series is so familiar to statisticians and applied mathematicians that it is worth stressing its limitations to emphasize that, invaluable though it may often be, neither it nor any other basis should be used uncritically. A Fourier series is especially useful for extremely stable functions, meaning functions where there are no strong local features and where the curvature tends to be of the same order everywhere. Ideally, the periodicity of the Fourier series should be reflected to some degree in the data, as is certainly the case for the temperature and gait data. Fourier series generally yield expansions which are uniformly smooth. But they are inappropriate to some degree for data known or suspected to reflect discontinuities in the function itself or in low-order derivatives. A Fourier series is like margarine: It is cheap and you can spread it on practically anything, but don't expect that the result will be exciting eating. Nevertheless, we find many applications for Fourier series expansion in this book.

3.2.4 *Polynomial bases*

The monomial basis $\phi_k(t) = (t - \omega)^k, k = 0, \ldots, K$ is also classic. Unfortunately, it can yield a nearly singular cross-product matrix $\Phi'\Phi$, and the shift parameter ω must be carefully chosen. However, if the argument values t_j are equally spaced or can be chosen to exhibit a few standard patterns, orthogonal polynomial expansions can be obtained, implying $O[(n+m)K]$ operations for all smooth values. Otherwise we are condemned to contemplate $O[(n+m)K^2]$ operations.

Like the Fourier series expansion, polynomials cannot exhibit very local features without using a large K. Moreover, polynomials tend to fit well in the centre of the data but exhibit rather unattractive behaviour in the tails. They are usually a poor basis for extrapolation or forecasting, for example.

Although derivatives of polynomial expansions are simple to compute, they are seldom satisfactory as estimators of the true derivative because of the rapid localized oscillation typical of high-order polynomial fits.

3.2.5 *Regression spline bases*

It is safe to say that the Fourier and polynomial bases have tended to be rather overused in applied work. Their inability to accommodate local features led to the development of polynomial splines: functions constructed by joining polynomials together smoothly at values τ_k called *knots*. Let the number of these knots be indicated by K_1+1, where the two outside knots define the interval $\mathcal{T}$ over which the estimation is to take place, and $K_1 - 1$ is the number of interior knots. Between two adjacent knots, a polynomial spline is a polynomial of fixed degree K_2, but at an interior knot, two adjacent polynomials are required to match in the values of a fixed number of their derivatives, usually chosen to be $K_2 - 1$. This implies that a spline of degree 0 is a step function discontinuous at knots, a spline of degree 1 is a polygonal or piecewise linear function, and a spline of degree 2 is a piecewise quadratic with continuous first derivative. A common choice of degree is 3, or piecewise cubic. The continuity of the first two derivatives of a cubic spline means that the curve is visually smooth. A classic reference on polynomial splines, in general, and B-splines, in particular, is de Boor (1978), and more recent references are Eubank (1988) and Green and Silverman (1994).

Polynomial splines combine the easy computability of polynomials with the capacity for changing local behaviour and great flexibility. There are many ways to represent polynomial splines, but perhaps the simplest conceptually is as a linear combination of the basis functions

$$\phi_k(t) = (t - \tau_k)_+^{K_2},$$

where u_+ means u if $u \geq 0$ and 0 otherwise, and where only the $K_1 - 1$ interior knot values are used. This is called the *truncated power basis.* This basis must be augmented by the monomial basis $t^k, k = 0, \ldots, K_2$. to produce a complete polynomial spline of order K. The total number of basis functions is $K = K_1 + K_2$, or the degree of the piecewise polynomial plus the number of interior knots.

Although the truncated power basis is still used in certain applications, its tendency to produce nearly singular cross-product matrices is a problem if there is more than a small number of knots. It is much better in practice to use *B-splines,* which have *compact support,* that is, are zero everywhere except over a finite interval. In the case of cubic splines, each B-spline is a cubic spline with support on the interval $[\tau_{k-2}, \tau_{k+2}]$, and with shorter support at the ends. Compact support implies both a band-structured cross-product matrix and $O(K)$ computation of all smoothing values.

There is a question as to where the knots τ_ℓ should be positioned. Although some applications strongly suggest specific knot locations, the choice may often be rather arbitrary, and users may complain that there is comparatively little to go on. On the other hand, some approaches capitalize on the free choice of knot location by beginning with a dense set of knots and then eliminating unneeded knots by an algorithmic procedure similar to variable selection techniques used in multiple regression. See, for example, Friedman and Silverman (1989).

3.2.6 *Wavelet bases*

We can construct a basis for all functions on $(-\infty, \infty)$ that are square-integrable by choosing a suitable *mother wavelet* function ψ and then considering all dilations and translations of the form

$$\psi_{jk}(t) = 2^{j/2}\psi(2^j t - k)$$

for integers j and k. We construct the mother wavelet to ensure that the basis is orthogonal, in the sense that the integral of the product of any two distinct basis functions is zero. Typically, the mother wavelet has compact support, and hence so do all the basis functions. The wavelet basis idea is easily adapted to deal with functions defined on a bounded interval, most simply if periodic boundary conditions are imposed.

The wavelet expansion of a function f gives a *multiresolution analysis* in the sense that the coefficient of ψ_{jk} yields information about f near position $2^{-j}k$ on scale 2^{-j}, i.e. at frequencies near $c2^j$ for some constant c. Thus, wavelets provide a systematic sequence of degrees of locality. In contrast to Fourier series, wavelet expansions cope well with discontinuities or rapid changes in behaviour; only those basis functions whose support includes the region of discontinuity or other bad behaviour are affected. This property, as well as a number of more technical mathematical results, means that it is often reasonable to assume that an observed function is well approximated by an economical wavelet expansion with few nonzero coefficients, even if it displays sharp local features.

Suppose a function x is observed without error at $n = 2^M$ regularly spaced points on an interval $\mathcal{T}$. Just as with the Fourier transformation, there is a discrete wavelet transform (DWT) which provides n coefficients closely related to the wavelet coefficients of the function x. We can calculate the DWT and its inverse in $O(n)$ operations, even faster than the $O(n \log n)$ of the FFT. As a consequence, most estimators based on wavelets can be computed extremely quickly, many of them in $O(n)$ operations.

Now suppose that the observations of x are subject to noise. The fact that many intuitively attractive classes of functions have economical wavelet expansions leads to a simple *nonlinear* smoothing approach: Construct the DWT of the noisy observations, and threshold it by throwing away the small coefficients in the expansion and possibly shrinking the large ones. The basic motivation of thresholding is the notion that any small coefficient is entirely noise and does not reflect any signal at all. This nonlinear thresholding has attractive and promising theoretical properties (see, for example, Donoho, Johnstone, Kerkyacharian and Picard, 1995), indicating that thresholded wavelet estimators should adapt well to different degrees of smoothness and regularity in the function being estimated.

The development of wavelet bases is comparatively recent, and only limited methodological experience is available at the time of writing. They are obviously of considerable promise and have excellent computational properties; perhaps their real practical strengths and weaknesses have yet to be fully understood. For further reading, see Chui (1992), Daubechies (1992), Press et al. (1992), Nason and Silverman (1994), Donoho et al. (1995), Johnstone and Silverman (1997), and the many references contained in these books and papers.

3.3 Smoothing by local weighting

3.3.1 Kernel functions

For a smoothing method to make any sense at all, the value of the function estimate at a point t must be influenced mostly by the observations near t. This feature is an implicit property of the estimators we have considered so far. In this section, we consider estimators where the local dependence is made more explicit by means of local weight functions.

The localizing weights w_j are simply constructed by a location and scale change of a *kernel* function with values $\text{Kern}(u)$. This kernel function is designed to have most of its mass concentrated close to 0, and either to decay rapidly or to disappear entirely for $|u| \geq 1$. Three commonly used kernels are

uniform:	$\text{Kern}(u) = 0.5, \ \ |u| \leq 1,$	0 otherwise ,
quadratic:	$\text{Kern}(u) = 0.75(1 - u^2), \ \ |u| \leq 1,$	0 otherwise , and
Gaussian:	$\text{Kern}(u) = (2\pi)^{-1/2} \exp(-u^2/2).$	

Then we define weight values

$$w_j(t) = \text{Kern}\left(\frac{t_j - t}{h}\right); \tag{3.11}$$

substantial values $w_j(t)$ as a function of j are now concentrated for t_j in the vicinity of t, and the degree of concentration is controlled by the size of h. The concentration parameter h is usually called the *bandwidth* parameter, and small values imply that only observations close to t receive any weight, whereas large h means that a wide-sweeping average uses values even a considerable distance from t.

3.3.2 Kernel smoothing

The simplest and classic case of an estimator that makes use of local weights is the *kernel estimator.* The estimate at a given point is a linear combination of local observations,

$$\hat{x}(t) = \sum_{j}^{N} S_j(t) y_j \tag{3.12}$$

for suitably defined weight functions S_j. Classically, the Nadaraya-Watson estimator (Nadaraya 1964, Watson 1964) is constructed by using the weights

$$S_j(t) = \frac{\mathtt{Kern}[(t_j - t)/h]}{\sum_r \mathtt{Kern}[(t_r - t)/h]}, \tag{3.13}$$

that is, the weight values $w_j(t)$ normalized to have a unit sum.

However, it is not essential that the smoothing weight values $S_j(t)$ sum exactly to one. The weights developed by Gasser and Müller (1979, 1984) are constructed as follows

$$S_j(t) = h \int_{\bar{t}_{j-1}}^{\bar{t}_j} \mathtt{Kern}\left(\frac{u - t}{h}\right) du \tag{3.14}$$

where $\bar{t}_j = (t_{j+1} + t_j)/2, 1 < j < n, \bar{t}_{-1} = t_1$ and $\bar{t}_n = t_n$. These weights are faster to compute, deal more sensibly with unequally spaced arguments, and have good asymptotic properties.

The need for fast computation strongly favours the compact support uniform and quadratic kernels, and the latter is the most efficient when only function values are required and the true underlying function x is twice-differentiable. The Gasser-Müller weights using the quadratic kernel are given by

$$S_j(t) = \frac{1}{4}[\{3r_{j-1}(t) - r_{j-1}^3(t)\} - \{3r_j(t) - r_j^3(t)\}]$$

for $|t_j - t| \le h$, and 0 otherwise, with

$$r_j(t) = \frac{t - \bar{t}_j}{h}. \tag{3.15}$$

We need to take special steps if t is within h units of either t_1 or t_n. These measures can consist of simply extending the data beyond this range in some reasonable way, making h progressively smaller as these limits are approached, or of sophisticated modifications of the basic kernel function Kern. The problem that all kernel smoothing algorithms

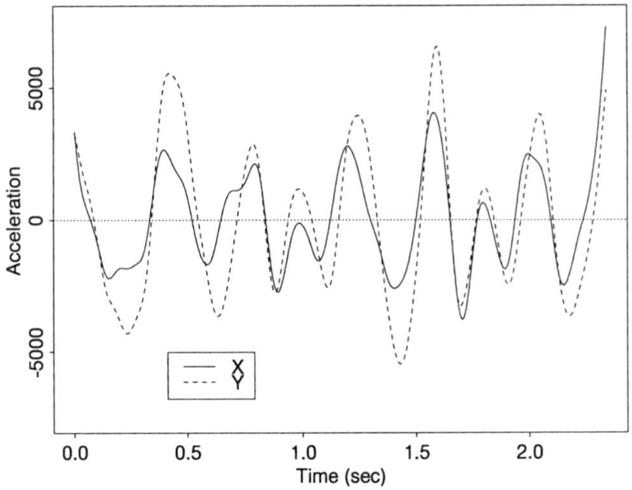

FIGURE 3.4. The second derivative or acceleration of the coordinate functions for the handwriting data. Kernel smoothing was used with a bandwidth $h = 0.075$.

have near the limits of the data is one of their major weaknesses, especially when h is large relative to the sampling rate.

Estimating the derivative just by taking the derivative of the kernel smooth is not usually a good idea, and, in any case, kernels such as the uniform and quadratic are not differentiable. However, kernels specifically designed to estimate a derivative of fixed order can be constructed by altering the nature of kernel function Kern. For example, a kernel Kern(u) suitable for estimating the first derivative must be zero near zero, positive above zero, and negative below, or it is a sort of smeared-out version of the first central difference. The Gasser-Müller weights for the estimation of the first derivative are

$$S_j(t) = \frac{15}{16h}[\{r_{j-1}^4(t) - 2r_{j-1}^2(t)\} - \{r_j^4(t) - 2r_j^2(t)\}] \tag{3.16}$$

and for the second derivative are

$$S_j(t) = \frac{105}{16h^2}[\{2r_{j-1}^3(t) - r_{j-1}^5(t) - r_{j-1}(t)\} - \{2r_j^3(t) - r_j^5(t) - r_j(t)\}] \tag{3.17}$$

for $|t_j - t| \leq h$ and 0 otherwise. It is usual to need a somewhat larger value of bandwidth h to estimate derivatives than is required for the function.

Figure 3.4 shows the estimated second derivative or acceleration for the two handwriting coordinate functions. After inspection of the results produced by a range of bandwidths, we settled on $h = 0.075$. This implies that any smoothed acceleration value is based on about 150 milliseconds of data and about 90 values of y_j.

3.3.3 Localized basis function estimators

The ideas of kernel estimators and basis function estimators can, in a sense, be combined to yield *localized basis function estimators*, which encompass a large class of function and derivative estimators. The basic idea is to extend the least squares criterion (3.6) to give a local measure of error as follows:

$$\mathsf{SMSSE}_t(y|c) = \sum_{j=1}^{n} w_j(t)[y_j - \sum_{k=1}^{K} c_k \phi_k(t_j)]^2, \tag{3.18}$$

where the weight functions w_j are constructed from the kernel function using (3.11).

In matrix terms,

$$\mathsf{SMSSE}_t(y|c) = (y - \Phi c)' \mathbf{W}(t)(y - \Phi c) \tag{3.19}$$

where $\mathbf{W}(t)$ is a diagonal matrix containing the weight values $w_j(t)$ in its diagonal. Choosing the coefficients c to minimize SMSSE_t yields

$$c = [\Phi' \mathbf{W}(t) \Phi]^{-1} \Phi' \mathbf{W}(t) y.$$

Substituting into the expansion $\hat{x}(t) = \sum_{k=1}^{K} c_k \phi_k(t)$ gives a linear smoothing estimator of the form (3.3) with smoothing weight values $S_j(t)$ the elements of the vector

$$S(t) = \mathbf{W}(t) \Phi [\Phi' \mathbf{W}(t) \Phi]^{-1} \phi(t) \tag{3.20}$$

where $\phi(t)$ is the vector with elements $\phi_k(t)$.

The weight values $w_j(t)$ in (3.18) are designed to have substantially nonzero values only for observations located close to the evaluation argument t at which the function is to be estimated. This implies that only the elements in $S(t)$ in (3.20) associated with data arguments values t_j close to the evaluation argument t are substantially different from zero. Consequently, $\hat{x}(t)$ is essentially a linear combination only of the observations y_j in the neighbourhood of t.

Since the basis has only to approximate a limited segment of the data surrounding t, the basis can do a better job of approximating the local features of the data and, at the same time, we can expect to do well with only a small number K of basis functions. The computational overhead for a single t depends on the number of data argument values t_j for which $w_j(t)$ is nonzero, as well as on K. Both of these are typically small. However, the price we pay for this flexibility is that the expansion must essentially be carried out anew for each evaluation point t.

3.3.4 Local polynomial smoothing

It is interesting to note that the Nadaraya-Watson kernel estimate can be obtained as a special case of the localized basis expansion method by setting $K = 1$ and $\phi_i(t) = 1$. A popular class of methods is obtained by extending from a single basis function to a low-order polynomial basis. Thus, we choose the estimated curve value $\hat{x}(t)$ to minimize the localized least squares criterion

$$\mathsf{SMSSE}_t(y|c) = \sum_{j=1}^{n} \mathsf{Kern}_h(t_j, t)[y_j - \sum_{\ell=0}^{L} c_\ell (t - t_j)^\ell]^2. \tag{3.21}$$

Setting $L = 0$, we recover the Nadaraya-Watson estimate. For values of $L \geq 1$, the function value and L of its derivatives can be estimated by the corresponding derivatives of the locally fitted polynomial at t. In general, the value of L should be at least one and preferably two higher than the highest order derivative required.

Local polynomial smoothing has a strong appeal; see, for example, the detailed discussion provided by Fan and Gijbels (1996). Its performance is superior in the region of the boundaries, and it adapts well to unequally spaced argument values. Local linear expansions give good results when we require only an estimate of the function value. They can easily be adapted in various ways to suit special requirements, such as robustness, monotonicity, and adaptive bandwidth selection. The S-PLUS function `lowess`, based on work of Cleveland (1979), incorporates a wide spectrum of valuable features.

3.3.5 Choosing the bandwidth

In all the localized basis expansion methods we have considered, the primary determinant of the degree of smoothing is the bandwidth h, rather than the number of basis functions used. The bandwidth controls the balance between two considerations, bias and variance in the estimate. Small values of h imply that the expected value of the estimate $\hat{x}(t)$ must be close to the true value $x(t)$, but the price we pay is in terms of the high variability of the estimate, since it is based on comparatively few observations. On the other hand, variability can always be decreased by increasing h, although this is inevitably at the expense of higher bias, since the values used cover a region in which the function's shape varies substantially. Mean squared error at t, which is the sum of squared bias and variance, provides a composite measure of performance.

There is a variety of data-driven automatic techniques for choosing an appropriate value of h, usually motivated by the need to minimize mean squared error across the estimated function. Unfortunately, none of these can always be trusted, and the problem of designing a reliable data-driven bandwidth selection algorithm continues to be a subject of active research and considerable controversy. Our own view is that trying out a variety of values of h and inspecting the consequences graphically remains a suitable means of resolving the bandwidth selection problem for most practical problems.

3.4 Summary

Explicitly localized smoothing methods, such as kernel smoothing and local polynomial smoothing, are easy to understand and have excellent computational characteristics. The role of the bandwidth parameter h is obvious, and as a consequence it is even possible to allow h to adapt to curvature variation. On the negative side, however, is the instability of these methods near the boundaries of the interval, although local polynomial smoothing performs substantially better than kernel smoothing in this regard. As with unweighted basis function expansions, it is well worthwhile to consider matching the choice of basis functions to known characteristics of the data, especially in regions where the data are sparse or where they are asymmetrically placed around the point t of interest, for example near the boundaries. The next chapter on the roughness penalty approach looks at the main competitor to kernel and local polynomial methods, namely spline smoothing.

4

The roughness penalty approach

4.1 Introduction

In this chapter we introduce a third option for approximating discrete data by a function. The roughness penalty or *regularization* approach retains the advantages of the basis function and local expansion smoothing techniques developed in Chapter 3, but circumvents some of their limitations. More importantly, it adapts gracefully to more general functional data analysis problems that we consider in subsequent chapters.

We saw that basis expansions work well if the basis functions have the same essential characteristics as the process generating the data. Thus, a Fourier basis is useful if the functions we observe are periodic and do not exhibit fluctuations in any particular interval that are much more rapid than those elsewhere. But basis expansions, on the debit side, have clumsy discontinuous control over the degree of smoothing, and can be expensive to compute if the basis exhibits neither orthogonality nor local support.

Standard kernel smoothing and local polynomial fitting techniques, on the other hand, are based on appealing, efficient and easily understood algorithms that are fairly simple modifications of classic statistical techniques. They offer continuous control of the smoothness of the approximation, but they are seldom optimal solutions to an explicit statistical problem, such as minimizing a measure of total squared error, and their rather heuristic character makes extending them to other smoothing situations difficult.

Like the basis expansion approach, roughness penalty methods are based on an explicit statement of what a smooth representation of the data is trying to do, but the need to have a smooth representation is expressed explicitly at the level of the criterion being optimized. Moreover, they can be applied to a much wider range of smoothing problems than simply estimating a curve x from observations of $x(t_i)$ for certain points t_i. Green and Silverman (1994) discuss a variety of statistical problems that can be approached using roughness penalties, including those where the data's dependence on the underlying curve is akin to the dependence on parameters in generalized linear models. Here we extend still further the scope of roughness penalty methods by discussing various functional data analysis contexts where roughness penalties are an elegant way to introduce smoothing into the analysis.

Finally, recent developments in the computational aspects of roughness penalty smoothing imply efficient $O(n)$ computation. These methods share the capacity to smooth a large number observations with kernel and local polynomial smoothing and with certain basis function expansions.

4.2 Spline smoothing

Let us first consider how regularization works in the simplest functional case when the goal is to estimate a function x on the basis of discrete and noisy observations in a vector y. Readers familiar with the multivariate linear model may wish to review the use of regularization for estimating a parameter matrix, developed in Section 2.6.

4.2.1 *Two competing objectives in function estimation*

The spline smoothing method estimates a curve x from observations $y_j = x(t_j) + \epsilon_j$ by making explicit two possible aims in curve estimation. On the one hand, we wish to ensure that the estimated curve gives a good fit to the data, for example in terms of the residual sum of squares $\sum_j \{y_j - x(t_j)\}^2$. On the other hand, we do not wish the fit to be too good if this results in a curve x that is excessively 'wiggly' or locally variable.

These competing aims can be seen, in some sense, as corresponding to the two elements of the basic principle of statistics:

$$\text{Mean squared error} = \text{Bias}^2 + \text{Sampling variance}$$

According to the error model $y_j = x(t_j) + \epsilon_j$, a completely unbiased estimate of the function value $x(t_j)$ can be produced by a curve fitting y_j exactly, since this observed value is itself an unbiassed estimate of $x(t_j)$. But any such curve must have high variance, manifested in the rapid local variation of the curve. In spline smoothing, as in other smoothing methods, the mean square error comes somewhat closer to capturing what we usually mean by poorness of estimation. It can often be dramatically reduced by sacrificing some bias to reduce sampling variance, and this is a key reason for imposing smoothness on the estimated curve. By requiring that the estimate vary only gently from one value to another, we are borrowing information from neighbouring data values, thereby expressing our faith in the regularity of the underlying function x that we are trying to estimate.

In the basis function approach, we constrained x to be a linear combination of a small number K of basis functions, and minimized the residual sum of squares subject to this hard-edged constraint. The philosophy of the roughness penalty method, on the other hand, is to allow a larger class of functions x but to quantify the rapid local variation of x and make an explicit trade-off between regularity and goodness of fit to the data, which corresponds to an implicit trade-off between variance and bias.

A popular measure of the roughness of a function is by its integrated squared second derivative

$$\mathrm{PEN}_2(x) = \int \{D^2 x(s)\}^2 \, ds = \|D^2 x\|^2. \tag{4.1}$$

This assesses the total curvature in x, or alternatively, the degree to which x departs from a straight line. Consequently, highly variable functions can be expected to yield high values of $\mathrm{PEN}_2(x)$ because their second derivatives are large over much of the range of interest.

We can then define the *penalized* residual sum of squares by

$$\mathrm{PENSSE}_\lambda(x|y) = \sum_j \{y_j - x(t_j)\}^2 + \lambda \times \mathrm{PEN}_2(x). \tag{4.2}$$

Our estimate of the function is obtained by finding the function x that minimizes $\mathrm{PENSSE}_\lambda(x)$ over the space of functions x for which $\mathrm{PEN}_2(x)$ is defined.

The parameter λ is a smoothing parameter that measures the rate of exchange between fit to the data, as measured by the residual sum of squares, and variability of the function x, as quantified by $\mathrm{PEN}_2(x)$. If λ is large, then functions which are not linear must incur a substantial roughness penalty in $\mathrm{PENSSE}_\lambda(x)$. For this reason, as $\lambda \to \infty$ the fitted curve x approaches the standard linear regression to the observed data.

On the other hand, for small λ the curve tends to become more and more variable since there is less and less penalty placed on its roughness, and as $\lambda \to 0$ the curve x approaches an interpolant to the data, satisfying $x(t_j) = y_j$ for all j. However, even in this limiting case the interpolating curve is not arbitrarily variable; instead, it is the smoothest twice-differentiable curve that exactly fits the data.

The curve x resulting from the use of the roughness penalty PEN_2 can be shown to be a cubic spline with knots at the data points t_j. Many details of the method are discussed by Green and Silverman (1994), and we discuss it only fairly superficially here, leaving a more detailed discussion to our treatment of a wider class of splines in Chapter 15. Function x can be found in $O(n)$ operations, for example using the S-PLUS function `smooth.spline`, which also contains options for choosing the smoothing parameter automatically from the data.

4.2.2 Estimating derivatives by spline smoothing

Many functional data analyses call for the estimation of derivatives, either because these are of direct interest, or because they play a role in some other part of the analysis. The penalty (4.1) may not be suitable, since it controls curvature in x itself, and therefore only slope in the derivative Dx. It does not require the second derivative D^2x even to be continuous, let alone smooth in any sense.

If the derivative of order m is the highest required, one should probably penalize two derivatives higher. For example, the estimate of acceleration is better if we use

$$\text{PEN}_4(x) = \int_T \{D^4x(s)\}^2 ds = \|D^4x\|^2 \tag{4.3}$$

in (4.2) since this controls the curvature in D^2x. Ramsay (1996c) developed an $O(n)$ algorithm for the polynomial smoothing spline defined by penalizing $D^m x$, and installed an S-PLUS module called `Pspline` in the Statlib library of S-PLUS functions. Statlib is accessible on the Web at `http://lib.stat.cmu.edu`. The software is also available through the home page for this book described in Section 1.8.

4.2.3 More general measures of fit

There are aspects of the roughness penalty method that are really useful in our treatment of functional data analysis. For example, instead of quantifying fit to the data by the residual sum of squares, we can penalize *any* criterion of fit by a roughness penalty. For instance, we might have a model for the observed y_j for which the log likelihood of x can be written down. Subtracting $\lambda \times \text{PEN}_2(x)$ from the log likelihood and then finding the maximum allows smoothing to be introduced in a wide range of statistical problems, not merely those in which error is appropriately measured by a residual sum of squares. These extensions of the roughness penalty method are a major theme of Green and Silverman (1994).

In the functional data analysis context, we adopt this philosophy in considering functional versions of several multivariate techniques. The function estimated by these methods is expressed as the solution of a maximization (or minimization) problem based on the given data. For example, principal components are chosen to have maximum possible variance subject to certain constraints. By penalizing this variance using a roughness penalty term appropriately, the original aim of the analysis can be traded off against the need to control the roughness of the estimate. There are different ways of incorporating the roughness penalty according to the context, but the overall idea remains the same: Penalize whatever is the appropriate measure of goodness-of-fit to the data for the problem under consideration.

4.2.4 *More general roughness penalties*

The second extension of the roughness penalty method uses measures of roughness other than $\|D^2x\|^2$. We have already seen one reason for this in Section 4.2.2, where the estimation of derivatives of x was considered. However, even if the function itself is of primary interest, there are two related reasons for considering more general roughness penalties. On the one hand, we may wish the class of functions with zero roughness were wider than, or otherwise different from, those that are of the form $a + bt$. On the other hand, we may have in mind that, locally at least, curves x should ideally satisfy a particular differential equation, and we may wish to penalize departure from this.

We can achieve both of these goals by replacing the second derivative operator D^2 with a more general linear differential operator L, defined as

$$Lx = w_0 x + w_1 Dx + \ldots + w_{m-1} D^{m-1} x + D^m x,$$

where the weights w_j may be either constants or functions $w_j(t)$. Then we can define

$$\text{PEN}_L(x) = \|Lx\|^2,$$

the integral of the square of Lx.

For instance, if we were observing periodic data on an interval $[0, 2\pi]$ and there were some reason to suppose that simple harmonic motion was a natural approximate model for the data, then we could define $Lx = D^2x + x$. Or, as an alternative to prespecifying the differential operator, we can use observed functional data to *estimate* the operator L. These ideas are developed further in Chapters 14 and 15.

4.3 The regularized basis approach

We now turn to a more general approach, of which spline smoothing turns out to be a special case. So far we have used basis functions in two essentially different ways. In Section 3.2, we forced the function x to lie in a relatively low dimensional space, defined in terms of a suitable basis. On the other hand, in Section 3.3, we did not assume that the whole function was in the span of a particular basis, but rather we considered a local basis expansion at any given point. In this section, we allow the function to have a higher dimensional basis expansion, but use a roughness penalty in fitting the function to the observed data.

4.3.1 *Complementary bases*

To develop our approach, suppose that we have two sets of basis functions, $\phi_j, j = 1, \ldots, J$ and $\psi_k, k = 1, \ldots, K$, that complement one another. Let functions ϕ_j be small in number and chosen to give reasonable account of the large-scale features of the data. The complementary basis functions ψ_k will generally be much larger in number, and are designed to catch local and other features not representable by the ϕ_j. Assume that any function x of interest can be expressed in terms of the two bases as

$$x(s) = \sum_{j=1}^{J} d_j \phi_j(s) + \sum_{k=1}^{K} c_k \psi_k(s). \tag{4.4}$$

For example, for the Canadian temperature data, the first three Fourier series functions with $\omega = \pi/6$ would be a natural choice for the ϕ_j, setting $J = 3$ and letting the ϕ basis be the functions

$$1, \ \sin(\omega t), \ \cos(\omega t).$$

The appropriate choice for the ψ_k in this case would be the remaining K functions in an order $(J + K)$ Fourier series expansion. In the monthly temperature data case, they could be the remaining nine Fourier series terms needed to represent the data exactly. Usually, as in the Fourier case above, the bases $\{\phi_j\}$ and $\{\psi_k\}$ are mutually linearly independent, and the expansion is unique, but this is not entirely essential to our method.

4.3.2 *Specifying the roughness penalty*

Let us now develop a roughness penalty for x so that linear combinations of the ϕ_j are in effect completely smooth, in that they contribute nothing to the roughness penalty. Then the roughness penalty must depend only on the coefficients of the ψ_k. One way of motivating this choice is by thinking of x as the sum of two parts, an 'ultrasmooth' function $x_S = \sum_j d_j \phi_j$ and a function $x_R = \sum_k c_k \psi_k$. Therefore we seek a measure $\text{PEN}(x_R)$ of how rough, or in any other way important, we would consider the function x_R expressed solely in terms of the ψ_k. One possibility is simply to take the usual L_2 norm of x_R, defining

$$\text{PEN}_0(x_R) = \int_{\mathcal{T}} x_R(s)^2 ds = \int (c'\psi)^2 = \int_{\mathcal{T}} [\sum_{k=1}^{K} c_k \psi_k(s)]^2 \, ds.$$

Another possibility is to take a certain order of derivative of the expansion prior to squaring and integrating, just as we did for the function x itself in Section 4.2. For example, we might use

$$\text{PEN}_2(x_R) = \int (D^2 x_R)^2 = \int_{\mathcal{T}} [\sum_{k=1}^{K} c_k D^2 \psi_k(s)]^2 \, ds$$

to assess the importance of x_R in terms of its total curvature, as measured by its squared second derivative, or $\text{PEN}_4(x_R) = \int (c' D^4 \psi)^2$ to assess the curvature of its second derivative. More generally, we can use any linear differential operator L, defining

$$\text{PEN}_L(x_R) = \int (L x_R)^2 = \int_{\mathcal{T}} [\sum_{k=1}^{K} c_k L \psi_k(s)]^2 \, ds.$$

Of course, setting L as the identity operator or the second derivative operator yields PEN_0 and PEN_2 as special cases.

We can express these penalties in matrix terms as

$$\text{PEN}_L(x_R) = c'\mathbf{R}c$$

where the order K symmetric matrix $\mathbf{R}$ contains elements

$$\mathbf{R}_{kl} = \int_{\mathcal{T}} L\psi_k(s) L\psi_l(s) ds = \langle L\psi_k, L\psi_l \rangle.$$

If computing the integrals proves difficult, a simple numerical integration scheme, such as the trapezoidal rule applied to a fine mesh of argument values, usually suffices, and then we can also estimate derivatives numerically. Alternatively, we can specify **R** directly as any suitable symmetric non-negative definite matrix, without explicit reference to the roughness of the function x_R.

Now we consider a general function x of the form (4.4), and simply define the roughness of x as

$$\text{PEN}(x) = c'\mathbf{R}c.$$

To express the penalized sum of squares, we need to express the residual sum of squares in terms of the coefficient vectors d and c. Working just as in (3.6),

$$\sum_i \{y_i - x(t_i)\}^2 = \|y - \Phi d - \Psi c\|^2$$

where the $n \times K$ matrix Ψ has elements $\Psi_{ik} = \psi_k(t_i)$. We can now define the composite smoothing criterion

$$\text{PENSSE}_\lambda(x|y) = \|y - \Phi d - \Psi c\|^2 + \lambda c'\mathbf{R}c. \tag{4.5}$$

We can minimize this quadratic form in d and c to find the fitted curve x in terms of its expansion (4.4) as follows. The solution for d for any fixed value of c is given by

$$d = (\Phi'\Phi)^{-1}\Phi'(y - \Psi c) \tag{4.6}$$

and, consequently,

$$\Phi d = \mathbf{P}_\phi(y - \Psi c),$$

where the projection matrix $\mathbf{P}_\phi = \Phi(\Phi'\Phi)^{-1}\Phi'$. In words, the ϕ basis component of the fit is the conventional basis expansion of the residual vector $y - \Psi c$. Substitute this solution for d into PENSSE_λ and define the complementary projection $\mathbf{Q}_\phi = \mathbf{I} - \mathbf{P}_\phi$. Recalling that $\mathbf{Q}_\phi\mathbf{Q}_\phi = \mathbf{Q}_\phi$, we arrive at the equation

$$c = (\Psi'\mathbf{Q}_\phi\Psi + \lambda\mathbf{R})^{-1}\Psi'\mathbf{Q}_\phi y. \tag{4.7}$$

Note that these calculations are for expository purposes only, and that there are often more stable and economical ways of actually computing a roughness penalty smooth. The published literature and algorithms on roughness penalty smoothing provide further details.

4.3.3 Some properties of the estimates

The first term of (4.5) is identical in structure to the error sum of squares criterion $Q(c)$ defined in (3.6), except that both sets of basis functions are used in the expansion. The second term, however, modifies the basis function expansion problem by penalizing the roughness or size in some sense of the ψ-component of the expansion.

The size of the penalty on the ψ-component is controlled by the smoothing parameter λ. In the limit as $\lambda \to 0$, no penalty whatsoever is applied, and the estimates obtained by minimizing the criterion PENSSE_λ revert to those obtained by an ordinary basis expansion in the combined basis of ϕ_j and ψ_k. At the other extreme, when $\lambda \to \infty$, the penalty is so severe that ψ-contribution to the roughness penalty is forced to zero; if $\mathbf{R}$ is strictly positive-definite, we obtain the basis function estimate corresponding to the basis $\{\phi_j\}$ alone. If $\mathbf{R}$ is not strictly positive-definite, then a contribution x_R from the $\{\psi_k\}$ basis is allowed, provided that it satisfies $Lx_R(s) = 0$ for all s.

It is instructive to study the minimizing values of the coefficient vectors d and c. The smoothing matrix then becomes

$$\mathbf{S}_\lambda = \mathbf{P}_\phi + \mathbf{Q}_{\psi,\lambda}$$

where

$$\mathbf{Q}_{\psi,\lambda} = \Psi(\Psi'\mathbf{Q}_\phi\Psi + \lambda\mathbf{R})^{-1}\Psi'\mathbf{Q}_\phi. \tag{4.8}$$

From this we can see that $\mathbf{Q}_{\psi,\lambda}$ is a kind of 'subprojection' matrix in the metric of the projection $\mathbf{Q}_\phi$ in that it has the structure of a true projection except for a perturbation of $\Psi'\mathbf{Q}_\phi\Psi$ by $\lambda\mathbf{R}$. This elaborates the way in which the regularized basis approach provides a continuous range of choices between low-dimensional basis expansion in terms of the functions ϕ_j and a high-dimensional expansion also making use of the functions ψ_k.

4.3.4 Relationship to the roughness penalty approach

We conclude with some remarks about the connections between the regularized basis method and the method discussed in Section 4.2.4 above. Firstly, to minimize the residual sum of squares penalized by $\|Lx\|^2$, we need not specify any functions at all in the ϕ_j part of the basis, but merely ensure that $\{\psi_k\}$ is a suitable basis for the functions of interest. In the original spline smoothing context, with $L = D^2$, we can take the $\{\psi_k\}$ to be a B-spline basis with knots at the data points, and, by using suitable methods of numerical linear algebra, we can obtain a stable $O(n)$ algorithm for spline smoothing; this is the approach of the S-PLUS function `smooth.spline`.

Secondly, if we wish to prescribe a particular ultrasmooth class $\mathcal{F}_0$, the regularized basis approach allows us to choose basis functions ϕ_j to span $\mathcal{F}_0$, and then allow $\mathbf{R}$ to be any appropriate strictly positive-definite matrix. In this way, the choice of the ultrasmooth class is decoupled from the way that roughness is measured.

5
The registration and display of functional data

5.1 Introduction

We can now assume that our observations are in functional form, and proceed to consider methods for their analysis. In this chapter, we consider aspects of the initial processing and display of functional data that raise issues specific to the functional context. Our main emphasis is on *registration* of the data, taking into account transformations of observed functions x that involve changes in the argument t as well as in the values $x(t)$ themselves. In the latter part of the chapter, we consider various descriptive statistics and graphical displays of functional data.

Figure 1.2 illustrates a problem that can frustrate even the simplest analyses of replicated curves. Ten records of the acceleration in children's height show, individually, the salient features of growth: The large deceleration during infancy is followed by a rather complex but small-sized acceleration phase during late childhood. Then the dramatic acceleration–deceleration pulses of the pubertal growth spurt finally give way to zero acceleration in adulthood. But the timing of these salient features obviously varies from child to child, and ignoring this timing variation in computing a cross-sectional mean function, shown by the heavy dashed line in Figure 1.2, can result in an estimate of average acceleration that does not resemble any of the observed curves. In this case, the mean curve has less variation during the pubertal phase than any single curve, and the duration of the mean pubertal growth spurt is rather larger than that of any individual curve.

The need to transform curves by transforming their arguments, which we call *curve registration*, can be motivated as follows. The rigid metric of physical time may not be directly relevant to the internal dynamics of many real-life systems. Rather, there can be a sort of biological or meteorological time scale that can be nonlinearly related to physical time, and can vary from case to case. Human growth, for example, is the consequence of a complex sequence of hormonal events that do not happen at the same rate for every child. Similarly, weather is driven by ocean currents, reflectance changes in land surfaces, and other factors timed differently for different spatial locations.

Put more abstractly, the values of two or more function values $x_i(t_j)$ can in principle differ because of two types of variation. The first is the more familiar vertical variation, or *range variation*, caused by the fact that $x_1(t)$ and $x_2(t)$ may simply differ at points of time t at which they are compared. But they may also exhibit *domain variation* in the sense that functions x_1 and x_2 should not be compared at the same time t. Instead, to compare the two functions, the time scale itself has to be distorted or transformed. For instance, the intensity of the pubertal growth spurts of two children should be compared at their respective ages of peak velocity rather than at any fixed age.

We now look at several types of curve registration problems, beginning with the problem of simply translating or shifting the values of t by a constant amount δ. Then we discuss landmark registration, which involves transforming t nonlinearly to line up important features or landmarks for all curves. Finally, we look at a more general method for curve registration.

5.2 Shift registration

5.2.1 *Types of alignment*

The pinch force data illustrated in Figure 1.7 are an example of a set of functional observations that must be aligned by shifting each horizontally before any meaningful cross-curve analysis is possible. This can happen because the time when the recording process begins is arbitrary, and is unrelated to the beginning of the interesting segment of the data, the period over which the measured squeeze actually takes place.

Let the interval $\mathcal{T}$ over which the functions are to be registered be $[T_1, T_2]$. We also need to assume that each sample function x_i is available for some region beyond each end of $\mathcal{T}$. The pinch force data, for example, are observed for substantial periods both before and after the force pulse that we wish to study. In the case of periodic data such as the Canadian temperature records, this requirement is easily met since one can wrap the function around by using the function's behaviour at the opposing end of the interval.

We are actually interested in the values

$$x_i^*(t) = x_i(t + \delta_i),$$

where the shift parameter δ_i is chosen to align the curves appropriately. For the pinch force data, the size of δ_i is of no real interest, since it merely measures the gap between the initialization of recording and the beginning of a squeeze. Silverman (1995) refers to this situation, in which a shift parameter must be accounted for but is of no real interest, as a *nuisance effects* problem.

The Canadian temperature data present a curve alignment problem of a somewhat different nature. As Figure 5.1 indicates, two temperature records, such as those for St. John's, Newfoundland, and Edmonton, Alberta, can differ noticeably in terms of the phase or timing of key events, such as the lowest mean temperature and the timing of spring and autumn. In this case, the shifts that would align these two curves vertically are of intrinsic interest, and should be viewed as a component of variation that needs careful description. It turns out that continental stations such as Edmonton have seasons earlier than marine stations such as St John's, because of the capacity of oceans to store heat and to release it slowly. In fact, either station's weather would have to be shifted by about three weeks to align the two.

When, as in the temperature data case, the shift is an important feature of each curve, we characterize its estimation as a *random effects* problem. Silverman (1995) also distinguishes a third and intermediate *fixed effects* case in which the shift must be carried out initially, and although not discarded completely once the the functions x_i^* have been constructed, is nevertheless only of tangential interest.

5.2.2 *Estimating the alignment*

The basic mechanics of estimating the shifts δ_i are the same, whether they are considered as nuisance or random effects. The differences become important when we consider the analysis in subsequent chapters, because in the random effects case (and, to some extent,

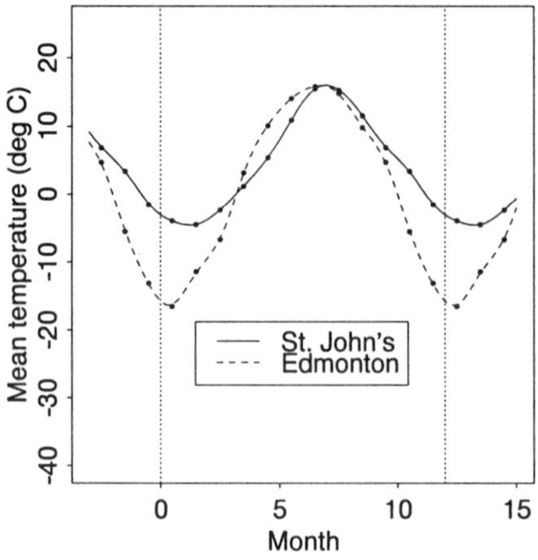

FIGURE 5.1. Temperature records for two weather stations where the timing of the seasons differs.

the fixed effects case) the δ_i enter the analysis. However, for present purposes we concentrate on the pinch force data as an example.

Estimating a shift or an alignment requires a criterion that defines when several curves are properly registered. One possibility is to identify a specific feature or *landmark* for a curve, and shift each curve so that this feature occurs at a fixed time. The time of the maximum of the smoothed pinch force is an obvious landmark. Note that this might also be expressed as the time when the first derivative crosses zero with negative slope. Landmarks are often more easily identifiable at the level of some derivative.

However, registration by landmark or feature alignment has some potentially undesirable aspects. The location of the feature may be ambiguous for certain curves, and if the alignment is only of a single point, variations in other regions may be ignored. For example, if we were to register the two temperature curves by aligning the midsummers, the midwinters might still remain seriously out of phase.

Instead, we can define a global registration criterion for identifying shift δ_i for curve i as follows. First we estimate an overall mean function $\hat{\mu}(s)$ for s in $\mathcal{T}$. If the individual functional observations x_i are smooth, it usually suffices to estimate $\hat{\mu}$ by the sample average $\bar{x}$. However, we wish to be able to evaluate derivatives of $\hat{\mu}$, and so more generally we want to smooth the overall estimate using one of the methods described in Chapters 3 and 4. We can now define our global registration criterion by

$$\begin{aligned}\text{REGSSE} &= \sum_{i=1}^{N} \int_{\mathcal{T}} [x_i(s+\delta_i) - \hat{\mu}(s)]^2\, ds \\ &= \sum_{i=1}^{N} \int_{\mathcal{T}} [x_i^*(s) - \hat{\mu}(s)]^2\, ds. \qquad (5.1)\end{aligned}$$

Thus, our measure of curve alignment is the integrated or global sum of squared vertical discrepancies between the shifted curves and the sample mean curve.

The target function for transformation in (5.1) is the unregistered cross-sectional estimated mean $\hat{\mu}$. But of course one of the goals of registration is to produce a better estimate of this same mean function. Therefore we expect to proceed iteratively. Beginning with the unregistered cross-sectional estimated mean, argument values for each curve are shifted so as to minimize REGSSE, then the estimated mean $\hat{\mu}$ is updated by re-estimating it from the *registered* curves x_i^*, and a new iteration is then undertaken using this revised target. This procedure of estimating a transformation by transforming to an iteratively updated average is often referred to as the *Procrustes method.* In practice, we have found that the process usually converges within one or two iterations.

A change of variable for each integration results in an alternative version of the criterion that may be more convenient for computational purposes:

$$\text{REGSSE} = \sum_{i=1}^{N} \int_{T_1+\delta_i}^{T_2+\delta_i} [x_i(s) - \hat{\mu}(s-\delta_i)]^2\, ds. \qquad (5.2)$$

5.2.3 Using the Newton-Raphson algorithm

We can estimate a specific shift parameter δ_i iteratively by using a modified Newton-Raphson algorithm for minimizing REGSSE. This procedure requires derivatives of REGSSE with respect to the δ_i. If we assume that the differences between x_i^* and $\hat{\mu}$ at the ends of the interval can be ignored (this is exactly true in the periodic case, and often approximately true in the nonperiodic case if the effects of real interest are concentrated in the middle of the interval), then

$$\frac{\partial}{\partial \delta_i}\text{REGSSE} = -\int_{T_1+\delta_i}^{T_2+\delta_i} \{x_i(s) - \hat{\mu}(s-\delta_i)\} D\hat{\mu}(s-\delta_i)\, ds$$

$$= \int_{\mathcal{T}} \{\hat{\mu}(s) - x_i^*(s)\} D\hat{\mu}(s)\, ds$$

and

$$\frac{\partial^2}{\partial \delta_i^2}\mathsf{REGSSE} = -\int_{\mathcal{T}} \{x_i^*(s) - \hat{\mu}(s)\} D^2\hat{\mu}(s)\, ds + \int_{\mathcal{T}} \{D\hat{\mu}(s)\}^2\, ds. \tag{5.3}$$

The modified Newton-Raphson algorithm works as follows:

Step 0: Begin with some initial shift estimates $\delta_i^{(0)}$, perhaps by aligning with respect to some feature, or even $\delta_i^{(0)} = 0$. But the better the initial estimate, the faster and more reliably the algorithm converges. Complete this step by estimating the average $\hat{\mu}$ of the shifted curves, using a method that allows the first two derivatives of $\hat{\mu}$ to give good estimates of the corresponding derivatives of the population mean, such as local polynomial regression of degree four, or roughness penalty smoothing with an integrated squared fourth derivative penalty.

Step ν, for $\nu = 1, 2, \ldots$: Modify the estimate $\delta_i^{(\nu-1)}$ on the previous iteration by

$$\delta_i^{(\nu)} = \delta_i^{(\nu-1)} - \alpha \frac{(\partial/\partial\delta_i)\mathsf{REGSSE}}{(\partial^2/\partial\delta_i^2)\mathsf{REGSSE}}$$

where α is a step-size parameter that can sometimes simply be set to one. It is usual to drop the first term (5.3) in the second derivative of REGSSE since it vanishes at the minimizing values, and convergence without this term tends to be more reliable when current estimates are substantially far from the minimizing values. Once the new shifts are estimated, recompute the estimated average $\hat{\mu}$ of the shifted curves.

Although the algorithm can in principle be iterated to convergence, and although convergence is generally fast, we have found that a single iteration is often sufficient with reasonable initial estimates. For the pinch force data, we initially aligned the smoothed curves by locating the maximum of each curve at 0.1 seconds. The shifts involved ranged from -20 to 50 milliseconds. We then carried out a single Newton-Raphson update ($\nu = 1$ above) where the range $\mathcal{T}$ of integration was from 23 to 251 milliseconds. The changes in the δ_i ranged from -3 to 2 milliseconds, and after this update, a second iteration did not yield

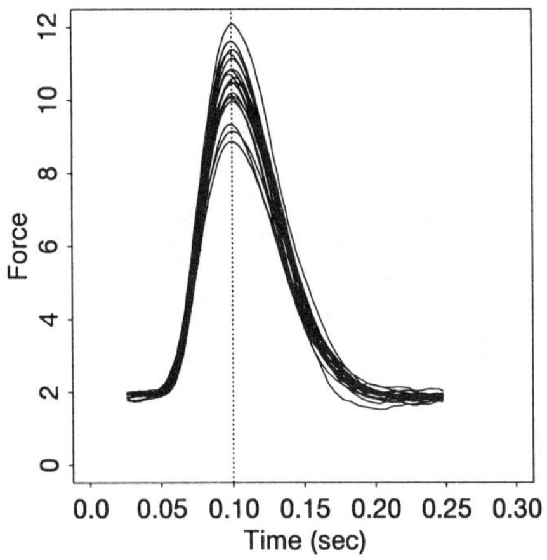

FIGURE 5.2. The pinch force curves aligned by minimizing the Procrustes criterion REGSSE.

any changes larger than a millisecond. The aligned curves are shown in Figure 5.2.

Kneip and Engel (1988) use linear transformations of t, both shift and scale changes, as part of a technique termed *self-modeling nonlinear regression* that attempts to estimate both parametric and nonparametric components of variation among several curves. Kneip and Engel (1995) use such shift-scale transformations to identify 'shape-invariant features' of curves, which remain unaltered by these changes in t.

5.3 Feature or landmark registration

A landmark or a feature of a curve is some characteristic that one can associate with a specific argument value t. These are typically maxima, minima, or zero crossings of curves, and may be identified at the level of some derivatives as well as at the level of the curves themselves.

We now turn to the more general problem of estimating a possibly nonlinear transformation h_i of t, and indicate how we can use landmarks to estimate this transformation. Coincidentally, the illustrative example we use shows how vector-valued functional data

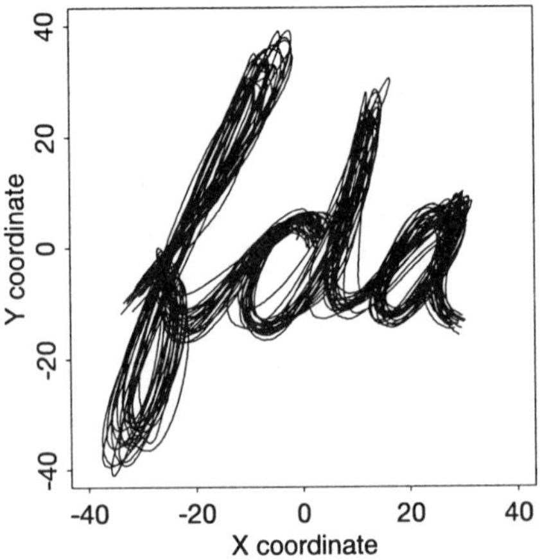

FIGURE 5.3. Twenty replications of "fda" written by one of the authors.

can be handled by obvious extensions of methods for scalar-valued functions.

The landmark registration process requires, for each curve x_i, the identification of the argument values $t_{if}, f = 1, \dots, F$ associated with each of F features. The goal is to construct a transformation h_i for each curve such that the registered curves with values

$$x^*(t) = x_i[h_i(t)]$$

have more or less identical argument values for any given landmark.

For example, consider the 20 replications of the letters "fda" in Figure 5.3. Each sample of handwriting was obtained by recording the position of a pen at a sampling rate of 600 times per second. There was some preprocessing to make each script begin and end at times 0 and 2.3 seconds, and to compute coordinates at the same 1,401 equally-spaced times. Each curve x_i in this situation is vector-valued, since two spatial coordinates are involved, and we use $\mathtt{ScriptX}_i$ and $\mathtt{ScriptY}_i$ to designate the X- and Y-coordinates, respectively.

Not surprisingly, there is some variation from observation to observation, and one goal is to explore the nature of this variation. But we want to take into account that variation in the "f," for example, can be of two sorts. There is temporal variation because timing of the top of the upper loop, for example, is variable. Although this type of variation would not show up in the plots in Figure 5.3, it may still be

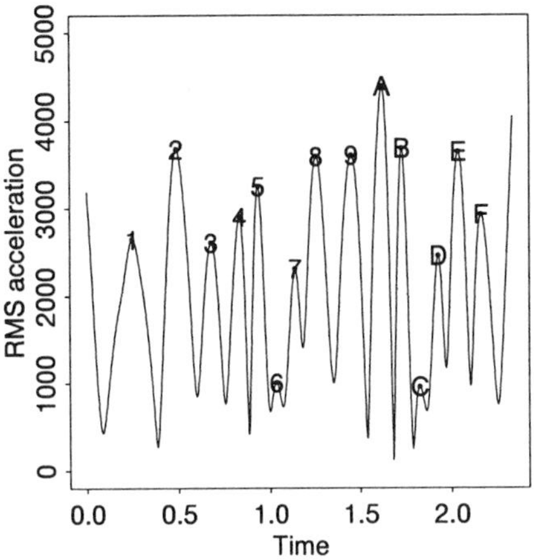

FIGURE 5.4. The average length of the acceleration vector for the 20 handwriting samples. The characters identify the 15 features used for landmark registration.

an important aspect of the way these curves vary. On the other hand, there is variation in the way the shape of each letter is formed, and this is obvious in the figure.

We estimated the accelerations or second derivatives of the two coordinate functions $D^2\mathtt{ScriptX}_i$ and $D^2\mathtt{ScriptY}_i$ by the local polynomial method described in Chapter 3. Figure 5.4 displays the average length of the acceleration vector

$$\sqrt{(D^2\mathtt{ScriptX}_i)^2 + (D^2\mathtt{ScriptY}_i)^2}$$

and we note that there are 15 clearly identified maxima, indicating points where the pen is changing direction. We also found that these maxima were easily identifiable in each record, and we were able to determine the values of t_{if} corresponding to them by just clicking on the appropriate points in a plot produced by S-PLUS. Figure 5.5 shows the first curve with these 15 features labelled, and we can see that landmarks labelled "4" and "A" mark the boundaries between letters. Figure 5.6 plots the values of the landmark timings t_{if} against the corresponding timings t_{0f} for the mean function. We were interested to see that the variability of the landmark timings was rather larger for the initial landmarks than for the later ones, and we were surprised by how small the variability was for all of them.

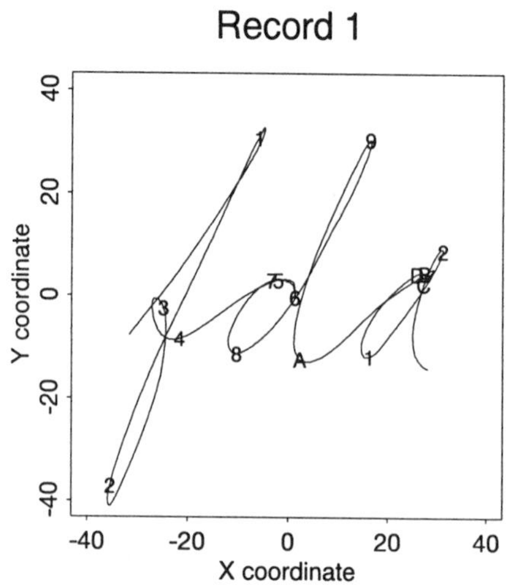

FIGURE 5.5. The first handwriting curve with the location of the 15 landmarks indicated by the characters used in Figure 5.4.

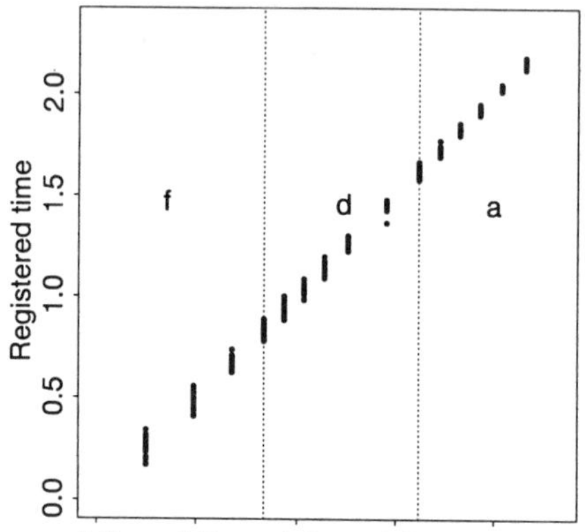

FIGURE 5.6. The timings of the landmarks for all 20 scripts plotted against the corresponding timings for the mean curve.

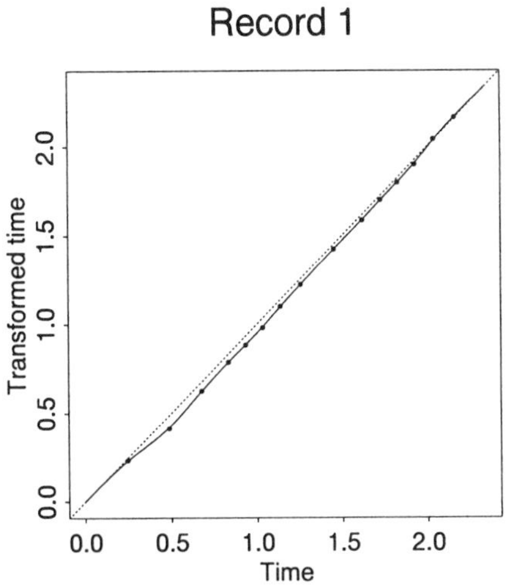

FIGURE 5.7. The time-warping function h_1 estimated for the first record that registers its features with respect to the mean curve.

The identification of landmarks enabled us to compare the X- and Y-coordinate values for the 20 curves at the landmark times, but of course we also wanted to make comparisons at arbitrary points between landmarks. This required the computation of a function h_i for each curve, called a *time-warping function* in the engineering literature, with the properties

- $h_i(0) = 0$
- $h_i(2.3) = 2.3$
- $h_i(t_{0f}) = t_{if}, f = 1, \ldots, 15$
- h_i is strictly monotonic: $s < t$ implies that $h_i(s) < h_i(t)$.

The values of the adjusted curves at time t are ScriptX[$h_i(t)$] and ScriptY[$h_i(t)$]. In all the adjusted curves, the landmarks each occur at the same time as in the mean function. In addition, the adjusted curves are also more or less aligned between landmarks. In this application, we merely used linear interpolation for time values between the points (t_{0f}, t_{if}) (as well as (0,0) and (2.3,2.3)) to define the time-warping function h_i for each curve. We introduce more sophisticated notions in the next section. Figure 5.7 shows the warping function computed in this manner for the first script record. Because h_1 is below the diagonal

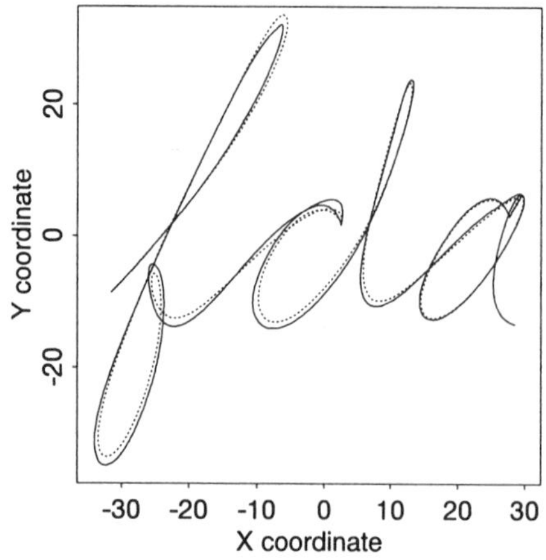

FIGURE 5.8. The solid line is the mean of the registered "fda" curves, and the dashed line is the mean of the unregistered curves.

line in the region of "f," the aligned time $h_1(t)$ is earlier than the actual time of features, and hence the actual times for curve 1 are retarded with respect to the mean curve.

We can now recompute the mean curve by averaging the registered curves. The result is in Figure 5.8, shown along with the mean for the unregistered data. Although the differences are not dramatic, as we might expect given the mild curvature in h_1, the upper and lower loops of the "f" are now more pronounced, and do represent the original curves substantially better.

Landmark registration has been studied in depth by Kneip and Gasser (1992), and their paper contains various technical details on the asymptotic behaviour of landmark estimates and warping functions estimated from them. They refer to the process of averaging a set of curves after registration as *structural averaging*, and their subsequent papers on growth curves (Gasser et al., 1990, 1991a,b) are important applications of this process.

For further information on the study of landmarks, the reader is referred to Gasser and Kneip (1995), who refer to a landmark as a 'structural feature', its location as a 'structural point', and to the distribution of landmark locations along the t axis as 'structural intensity'. They have developed some useful tools for estimating landmark distribution or structural intensity, and, indeed, the study of the characteristics and distribution of curve characteristics is an important part of functional data analysis in many applications. Another source of much information on the study of landmarks and their use in registration is Bookstein (1991).

5.4 More general transformations

5.4.1 *Families of transformations*

In this section, we consider more general transformations and draw together some of the ideas of Sections 5.2 and 5.3. Let an arbitrary transformation of argument t be indicated by $h(t)$. The function h has the requirements of being strictly increasing or monotonic, and also differentiable up to a certain order. The general registration problem is one of estimating such an h_i for each curve x_i, such that the registered curves

$$x_i^*(t) = x_i[h_i(t)]$$

satisfy some criterion, such as having aligned local features or minimizing a measure like REGSSE. Our discussion concentrates almost exclusively on the minimization of criteria related to REGSSE, but the ideas can be extended to other criteria of goodness of fit of a transformed functional observation to an overall mean.

The transformation family h may be *parametric* in character, an example of which is the translation $h_i(t) = t + \delta_i$, or the linear transformation $h_i(t) = (t + \delta_i)\beta_i, \beta_i > 0$. For each record i, let the vector γ_i contain the parameters, such as δ_i and β_i, defining the transformation, and let $h_i(t)$ indicate the value of the transformation for parameter vector value γ_i.

In this case, the Procrustes method is defined as the iterative minimization of

$$\text{REGSSE} = \sum_{i=1}^{N} \int_{\mathcal{T}} [x_i\{h_i(s)\} - \hat{\mu}(s)]^2 \, ds. \tag{5.4}$$

with $\hat{\mu}(s)$ re-estimated from the curves $x_i\{h_i(s)\}$ at each stage. The Newton–Raphson update of the parameter estimates is defined as above; the partial derivatives are

$$\frac{\partial}{\partial \gamma_{iq}} \text{REGSSE} = \int_{\mathcal{T}} [x_i\{h_i(s)\} - \hat{\mu}(s)] D x_i[h_i(s)] \frac{\partial h_i(s)}{\partial \gamma_{iq}} \, ds$$

and

$$\frac{\partial^2}{\partial \gamma_{iq}^2}\text{REGSSE} = \int_{\mathcal{T}} [x_i\{h_i(s)\} - \hat{\mu}(s)] D^2 x_i[h_i(s)] \frac{\partial^2 h_i(s)}{\partial \gamma_{iq}^2}\, ds \quad (5.5)$$
$$- \int_{\mathcal{T}} \{D x_i(h_i(t;\gamma_i)) \frac{\partial h_i(s)}{\partial \gamma_{iq}}\}^2\, ds.$$

The function $Dx_i[h_i(s)]$ is the derivative $dx_i(t)/dt$ evaluated at $t = h_i(s)$, and similarly for $D^2x_i[h_i(s)]$. These values may usually be estimated by a suitable interpolation or smoothing method. It is often numerically more stable to approximate the second derivative of REGSSE by dropping the first term on the right side of (5.5).

5.4.2 *Fitting transformations using regularization*

Ramsay (1996b) and Ramsay and Li (1996) have developed the fitting of a general and flexible family of warping functions h_i making use of a regularization technique. The results are sketched here; for further details see the original papers.

Suppose that the interval $\mathcal{T}$ of interest is $[0, T]$, and that the transformation functions h have integrable second derivative in addition to being strictly increasing. Every such function can be described by the homogeneous linear differential equation

$$D^2 h = wDh \quad (5.6)$$

for some suitable weight function w. To see this, note that a strictly monotonic function has a nonzero derivative, a twice-differentiable function has a second derivative, and therefore the weight function w is trivially D^2h/Dh. The weight function $w(s)$ is a measure of *relative curvature* of $h(s)$ in the sense that the conventional curvature measure $D^2h(s)$ is assessed relative to the slope Dh.

The differential equation (5.6) has the general solution

$$\begin{aligned} h(t) &= C_0 + C_1 \int_0^t [\exp \int_0^u w(v)\, dv]\, du \\ &= C_0 + C_1 (D^{-1} \exp D^{-1} w)(t) \end{aligned} \quad (5.7)$$

where D^{-1} is the integration operator (with lower limit of integration 0 and variable upper limit), and C_0 and C_1 are arbitrary constants. When w is constant, $h(t) = C_0 + C_1 \exp(wt)$, so that an exponential function has constant relative curvature in this sense. A straight line is of course implied by $w(s) = 0$ for all s. In general, we impose the constraints

$h(0) = 0$ and $h(T) = T$, implying that

$$C_0 = 0 \text{ and } C_1 = T/[D^{-1} \exp D^{-1} w(T)]$$

and that the function h depends only on the weight function w.

We can regularize or smooth the estimated warping functions h_i by augmenting the criterion (5.4) as follows:

$$\mathsf{REGSSE}_\lambda(h_1, \ldots, h_n) = \mathsf{REGSSE} + \lambda \sum_i \int w_i^2(t)\, dt, \tag{5.8}$$

where $w_i = D^2 h_i / D h_i$. Larger values of the smoothing parameter λ shrink the relative curvature functions w_i to zero, and therefore shrink the $h_i(t)$ to t.

5.4.3 *A regularized basis approach*

In principle, we could consider the minimization of REGSSE_λ without any further constraints on the functions h_i, but it is computationally much simpler to restrict attention to a particular high-dimensional family of transformations for which the minimization is tractable. To this end, we can use a linear B-spline basis for $D^{-1}w$.

Beginning with a set of discrete break points $\tau_k, k = 0, \ldots, K$ satisfying $0 = \tau_0 < \tau_1 < \ldots < \tau_K = T$, let

$$\Delta_k = \tau_{k+1} - \tau_k, \; k = 0, \ldots, K-1,$$

and for notational simplicity let Δ_k also indicate the interval $[\tau_k, \tau_{k+1})$. Define $\mathtt{Triang}_k$ to be the triangle or 'hat' function

$$\mathtt{Triang}_k(t) = \begin{cases} (t - \tau_{k-1})/\Delta_{k-1} & \text{if } t \in \Delta_{k-1} \\ (\tau_{k+1} - t)/\Delta_k & \text{if } t \in \Delta_k \\ 0 & \text{otherwise} \end{cases} \tag{5.9}$$

for $k = 1, \ldots, K$. (For $k = K$ there is no interval Δ_K to worry about.)

For any particular weight function w, let $c_k = D^{-1}w(\tau_k)$, implying $D^{-1}w = \sum_{k=1}^K c_k \mathtt{Triang}_k$. In each interval Δ_k, the weight function w takes the constant value $w_k = (c_{k+1} - c_k)/\Delta_k$, and so, if required, we can calculate the penalty

$$\int_0^T w(t)^2\, dt = \sum_{k=0}^{K-1} \Delta_k^{-1} (c_{k+1} - c_k)^2.$$

The integrals in (5.7) can be computed explicitly. For $t \in \Delta_k$

$$\begin{aligned} h(t) &= h(\tau_k) + C_1 w_k^{-1} \exp[c_k + w_k(t - \tau_k)] \text{ and} \\ Dh(t) &= C_1 \exp[c_k + w_k(t - \tau_k)], \end{aligned} \tag{5.10}$$

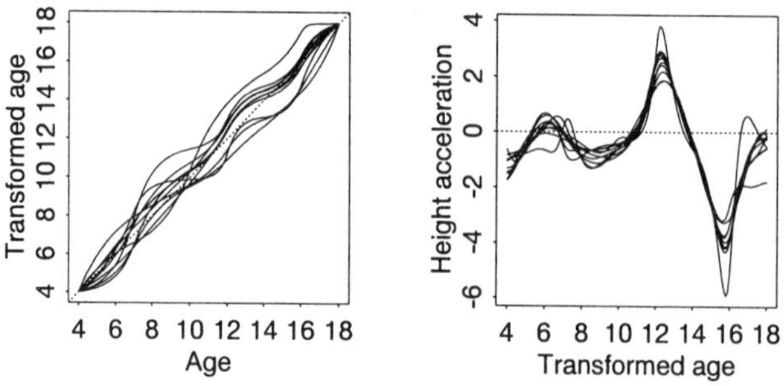

FIGURE 5.9. The left panel contains the estimated time warping functions h_i for the ten height acceleration curves in Figure 1.2. The right panel displays the registered curves.

where, as already indicated, C_1 is a normalizing constant yielding $h(T) = T$.

For any particular fitting criterion, we can estimate the coefficients c_k by a numerical procedure. Computation of the value t corresponding to $h(t) = z$ for a given z is often required, in other words the value of the inverse function $h^{-1}(z)$. Evaluation of h^{-1} is accomplished by first identifying the interval Δ_k such that $z = h(t), t \in \Delta_k$.

5.4.4 *Registering the height acceleration curves*

The ten acceleration functions in Figure 1.2 were registered by the Procrustes method and the regularized basis expansion method as set out in Section 5.4.3. The interval $\mathcal{T}$ was taken to be $[4, 18]$ with time measured in years. The break values τ_k defining the monotonic transformation family (5.10) were 4, 7, 10, 12, 14, 16 and 18 years, and the curves were registered over the interval [4,18] using criterion (5.8) with $\lambda = 0.001$. A single Procrustes iteration produced the results displayed in Figure 5.9. The left panel displays the ten warping functions h_i, and the right panel shows the curve values $x_i[h_i(t)]$. Figure 5.10 compares the unregistered and registered cross-sectional means. We see that the differences are substantial, and moreover that the mean of the registered function tends to resemble most of the sample curves much more closely.

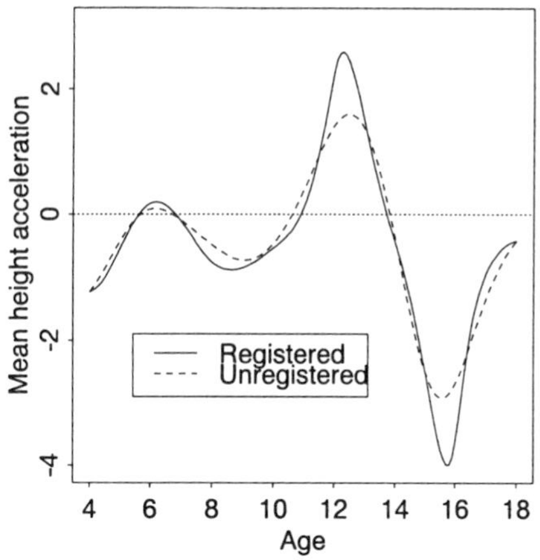

FIGURE 5.10. The cross-sectional means of the registered and unregistered height acceleration curves displayed in Figure 1.2.

5.4.5 Some cautionary remarks

In general it is not clear that variation in the amplitude of curves can be cleanly separated from the variation that the registration process aims to account for. It is easy to construct examples where a registration function h that is allowed to be highly nonlinear can remove variation that is clearly of an amplitude nature. This problem of lack of identifiability of the two types of variation, horizontal and vertical, is perhaps less of a concern if only linear transformations are permitted, and is also not acute for landmark registration, where the role of the transformation is only to align curve features.

But if flexible families of monotonic transformations such as those described above are used in conjunction with a global fitting criterion such as REGSSE, two pieces of practical advice are offered. First, allow transformations to differ from linear only with caution. Second, first remove amplitude effects that can be accounted for by vertical shifts or scale changes before registration, by centering and possibly rescaling curves. Alternatively, it may be wiser to register the first or second derivatives of curves, as we have done with the growth curves, rather than the curves themselves, because differences in centering are automatically removed when derivatives are taken.

6

Principal components analysis for functional data

6.1 Introduction

For many reasons, principal components analysis (PCA) of functional data is a key technique to consider. First, our own experience is that, after the preliminary steps of registering and displaying the data, the user wants to explore that data to see the features characterizing typical functions. Some of these features are expected to be there, for example the sinusoidal nature of temperature curves, but other aspects may be surprising. Some indication of the complexity of the data is also required, in the sense of how many types of curves and characteristics are to be found. Principal components analysis serves these ends admirably, and it is perhaps also for these reasons that it was the first method to be considered in the early literature on FDA.

Just as for the corresponding matrices in the classical multivariate case, the variance-covariance and correlation functions can be difficult to interpret, and do not always give a fully comprehensible presentation of the structure of the variability in the observed data directly. The same is true, of course, for variance-covariance and correlation matrices in classical multivariate analysis. A principal components analysis provides a way of looking at covariance structure that can be much more informative and can complement, or even replace altogether, a direct examination of the variance-covariance function.

PCA also offers an opportunity to consider some issues that reappear in subsequent chapters. For example, we consider immediately how PCA is defined by the notion of a linear combination of function values, and why this notion, at the heart of most of multivariate data analysis, requires some care in a functional context. A second issue is that of *regularization*; for many data sets, PCA of functional data is more revealing if some type of smoothness is required of the principal components themselves. We consider this topic in detail in Chapter 7.

6.2 Defining functional PCA

6.2.1 PCA for multivariate data

The central concept exploited over and over again in multivariate statistics is that of taking a linear combination of variable values,

$$f_i = \sum_{j=1}^{p} \beta_j x_{ij}, \quad i = 1, \dots, N \tag{6.1}$$

where β_j is a weighting coefficient applied to the observed values x_{ij} of the jth variable. We can express (6.1) somewhat more compactly by using the inner product notation introduced in Chapter 2:

$$f_i = \langle \beta, x_i \rangle, \quad i = 1, \dots, N \tag{6.2}$$

where β is the vector $(\beta_1, \dots, \beta_p)'$ and x_i is the vector $(x_{i1}, \dots, x_{ip})'$.

In the multivariate situation, we choose the weights so as to highlight or display types of variation that are very strongly represented in the data. Principal components analysis can be defined in terms of the following stepwise procedure, which defines sets of normalized weights that maximize variation in the f_i's:

1. Find the weight vector $\xi_1 = (\xi_{11}, \dots, \xi_{p1})'$ for which the linear combination values

$$f_{i1} = \sum_j \xi_{j1} x_{ij} = \langle \xi_1, x_i \rangle$$

have the largest possible mean square $N^{-1} \sum_i f_{i1}^2$ subject to the constraint

$$\sum_j \xi_{j1}^2 = \|\xi_1\|^2 = 1.$$

2. Carry out second and subsequent steps, possibly up to a limit of the number of variables p. On the mth step, compute a new weight vector ξ_m with components ξ_{jm} and new values $f_{im} = \langle \xi_m, x_i \rangle$. Thus, the values f_{im} have maximum mean square, subject to the constraint $\|\xi_m\|^2 = 1$ and the $m - 1$ additional constraint(s)

$$\sum_j \xi_{jk}\xi_{jm} = \langle \xi_k, \xi_m \rangle = 0, \; k < m.$$

The motivation for the first step is that by maximizing the mean square, we are identifying the strongest and most important mode of variation in the variables. The unit sum of squares constraint on the weights is essential to make the problem well defined; without it, the mean squares of the linear combination values could be made arbitrarily large. On second and subsequent steps, we seek the most important modes of variation again, but require the weights defining them to be orthogonal to those identified previously, so that they are indicating something new. Of course, the amount of variation measured in terms of $N^{-1} \sum_i f_{im}^2$ will decline on each step. At some point, usually well short of the maximum index p, we expect to lose interest in modes of variation thus defined.

The definition of principal components analysis does not actually specify the weights uniquely; for example, it is always possible to change the signs of all the values in any vector ξ_m without changing the value of the variance that it defines.

The values of the linear combinations f_{im} are called *principal component scores* and are often of great help in describing what these important components of variation mean in terms of the characteristics of specific cases or replicates.

To be sure, the mean is a very important aspect of the data, but we already have an easy technique for identifying it. Therefore, we usually subtract the mean for each variable from corresponding variable values before doing PCA. When this is done, maximizing the mean square of the principal component scores corresponds to maximizing their sample variance.

6.2.2 Defining PCA for functional data

How does PCA work in the functional context? The counterparts of variable values are function values $x_i(s)$, so that the discrete index j in the multivariate context has been replaced by the continuous index s. Summations over j are replaced by integrations over s to define the linear 'combinations'

$$f_i = \int \beta(s)x_i(s)\,ds = \langle \beta, x_i \rangle. \tag{6.3}$$

The weights β_j now become a weight function with values $\beta(s)$.

In the first functional PCA step, the weight function ξ_1 is chosen to maximize $N^{-1}\sum_i f_{i1}^2 = N^{-1}\sum_i \langle \xi_1, x_i \rangle^2$ subject to the continuous analogue of the unit sum of squares constraint $\|\xi_1\|^2 = 1$. This time, the notation $\|\xi_1\|^2$ is used to mean the squared norm $\int \xi_1(s)^2 ds$ of the *function* ξ_1.

Postponing computational details until Section 6.4, now consider as an illustration the upper left panel in Figure 6.1. This displays the weight function ξ_1 for the Canadian temperature data after the mean across all 35 weather stations has been removed from each station's monthly temperature record. Although ξ_1 is positive throughout the year, the weight placed on the winter temperatures is about four times that placed on summer temperatures. This means that the greatest variability between weather stations will be found by heavily weighting winter temperatures, with only a light contribution from the summer months; Canadian weather is most variable in the wintertime, in short. Moreover, the percentage 89.3% at the top of the panel indicates that this type of variation strongly dominates all other types of variation. Weather stations for which the score f_{i1} is high will have much warmer than average winters combined with warm summers, and the two highest scores are in fact assigned to Vancouver and Victoria on the Pacific Coast. To no one's surprise, the largest negative score goes to Resolute in the High Arctic.

As for multivariate PCA, the weight function ξ_m is also required to satisfy the orthogonality constraint(s) $\langle \xi_k, \xi_m \rangle = 0,\ k < m$ on subsequent steps. Each weight function has the task of defining the most important mode of variation in the curves subject to each mode being orthogonal to all modes defined on previous steps. Note again that the weight functions are defined only to within a sign change.

The weight function ξ_2 for the temperature data is displayed in the upper right panel of Figure 6.1. Because it must be orthogonal to ξ_1, we cannot expect that it will define a mode of variation in the temperature functions that will be as important as the first. In fact, this second mode accounts for only 8.3% of the total variation, and consists of a positive contribution for the winter months and a negative contribution for the summer months, therefore corresponding to a measure of uniformity of temperature through the year. On this component, one of the highest scores f_{i2} goes to Prince Rupert, also on the Pacific coast, for which there is comparatively low discrepancy between winter and summer.

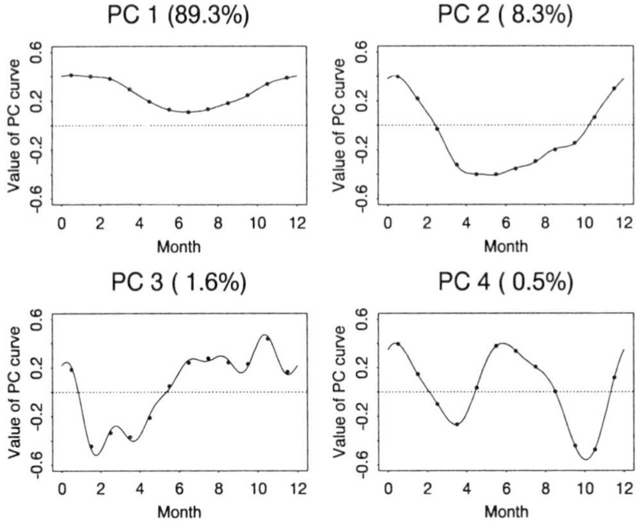

FIGURE 6.1. The first four principal component curves of the Canadian temperature data estimated by two techniques. The points are the estimates from the discretization approach, and the curves are the estimates from the expansion of the data in terms of a 12-term Fourier series. The percentages indicate the amount of total variation accounted for by each principal component.

Prairie stations such as Winnipeg, on the other hand, have hot summers and very cold winters, and receive large negative second component scores.

The third and fourth components account for small proportions of the variation, since they are required to be orthogonal to the first two as well as to each other. At this point they are difficult to interpret, but we look at techniques for understanding them in Section 6.3.

Displays such as Figure 6.1 can remind one of the diagrams of modes of vibration in a string fixed at both ends always found in introductory physics texts. The first and dominant type is simple in structure and resembles a single cycle of a sine wave. Subdominant or higher order components are also roughly sinusoidal, but with more and more cycles. With this analogy in mind, we find the term *harmonics* evocative in referring to principal components of variation in curves in general.

6.2.3 *Defining an optimal empirical orthonormal basis*

There are several other ways to motivate PCA, and one is to define the following problem: We want to find a set of exactly K orthonormal functions ξ_m so that the expansion of each curve in terms of these basis functions approximates the curve as closely as possible. Since these basis functions are orthonormal, it follows that the expansion will be of the form

$$\hat{x}_i(t) = \sum_{k=1}^{K} f_{ik}\xi_k(t)$$

where $f_{ik} = \langle x_i, \xi_k \rangle$. As a fitting criterion for an individual curve, consider the integrated squared error

$$\|x_i - \hat{x}_i\|^2 = \int [x(s) - \hat{x}(s)]^2 \, ds$$

and as a global measure of approximation

$$\text{PCASSE} = \sum_{i=1}^{N} \|x_i - \hat{x}_i\|^2. \tag{6.4}$$

The problem is then, more precisely, what choice of basis will minimize the error criterion (6.4)?

The answer, it turns out, is precisely the same set of principal component weight functions that maximize variance components as defined above. For this reason, these functions ξ_m are referred to in some fields as *empirical orthonormal functions*, because they are determined by the data they are used to expand.

6.2.4 *PCA and eigenanalysis*

In this section, we investigate another characterization of PCA, in terms of the eigenanalysis of the variance-covariance function or operator.

Assume for this section that our observed values, x_{ij} in the multivariate context and $x_i(t)$ in the functional situation, result from subtracting the mean variable or function values, so that their sample means $N^{-1}\sum_i x_{ij}$, or cross-sectional means $N^{-1}\sum_i x_i(t)$, respectively, are zero.

Texts on multivariate data analysis tend to define principal components analysis as the task of finding the eigenvalues and eigenvectors of the covariance or correlation matrix. The logic for this is as follows. Let the $N \times p$ matrix $\mathbf{X}$ contain the values x_{ij} and the vector ξ of length p contain the weights for a linear combination. Then the mean square criterion for finding the first principal component weight vector can be written as

$$\max_{\xi'\xi=1} N^{-1}\xi'\mathbf{X}'\mathbf{X}\xi \tag{6.5}$$

since the vector of principal component scores f_i can be written as $\mathbf{X}\xi$.

Use the $p \times p$ matrix $\mathbf{V}$ to indicate the sample variance-covariance matrix $\mathbf{V} = N^{-1}\mathbf{X}'\mathbf{X}$. (One may prefer to use a divisor of $N-1$ to N since the means have been estimated, but it makes no essential difference to the principal components analysis.) The criterion (6.5) can now be expressed as

$$\max_{\xi'\xi=1} \xi'\mathbf{V}\xi.$$

As explained in Section A.3, this maximization problem is now solved by finding the solution with largest eigenvalue ρ of the eigenvector problem or eigenequation

$$\mathbf{V}\xi = \rho\xi. \tag{6.6}$$

There is a sequence of different eigenvalue-eigenvector pairs (ρ_j, ξ_j) satisfying this equation, and the eigenvectors ξ_j are orthogonal. Because the mean of each column of $\mathbf{X}$ is usually subtracted from all values in that column as a preliminary to principal components analysis, the rank of $\mathbf{X}$ is $N-1$ at most, and hence the $p \times p$ matrix $\mathbf{V}$ has, at most, $\min\{p, N-1\}$ nonzero eigenvalues ρ_j. For each j, the eigenvector ξ_j satisfies the maximization problem (6.5) subject to the additional constraint of being orthogonal to all the eigenvectors $\xi_1, \xi_2, \ldots, \xi_{j-1}$ found so far. This is precisely what was required of the principal components in the second step laid out in Section 6.2.1. Therefore, as we have defined it, the multivariate PCA problem is equivalent to the algebraic and numerical problem of solving the eigenequation (6.6). Of course, there are standard computer algorithms for doing this.

Appealing to the more general results set out in Section A.3.2, the functional PCA problem leads to the eigenequation

$$\int v(s,t)\xi(t)\,dt = \langle v(s,\cdot), \xi\rangle = \rho\xi(s) \tag{6.7}$$

where the covariance function v is given by

$$v(s,t) = N^{-1}\sum_{i=1}^{N} x_i(s)x_i(t). \tag{6.8}$$

Again, note that we may prefer to use $N-1$ to define the variance-covariance function v; nothing discussed here changes in any essential way.

The left side of (6.7) is an *integral transform* V of the weight function ξ with the kernel of the transform v defined by

$$V\xi = \int v(\cdot, t)\xi(t)\, dt. \tag{6.9}$$

This integral transform is called the *covariance operator* V because the integral transform acts or operates on ξ. Therefore we may also express the eigenequation directly as

$$V\xi = \rho\xi,$$

where ξ is now an eigenfunction rather than an eigenvector.

One important difference between the multivariate and functional eigenanalysis problems concerns the maximum number of different eigenvalue-eigenfunction pairs. The counterpart of the number of variables p in the multivariate case is the number of function values in the functional case, and thus infinity. Provided the functions x_i are not linearly dependent, the operator V will have rank $N - 1$, and there will be $N - 1$ nonzero eigenvalues.

To summarize, in this section we find that principal components analysis is defined as the search for a set of mutually orthogonal and normalized weight functions ξ_m. Functional PCA can be expressed as the problem of the eigenanalysis of the covariance operator V, defined by using the covariance function v as the kernel of an integral transform. The use of inner product notation is helpful here because the same notation can be used whether the data are multivariate or functional.

In Section 6.4 we discuss practical methods for actually computing the eigenfunctions ξ_m, but first we consider some aspects of the display of principal components once they have been found.

6.3 Visualizing the results

The fact that interpreting the components is not always an entirely straightforward matter is common to most functional PCA problems. We now consider some techniques that may aid their interpretation.

6.3.1 Plotting components as perturbations of the mean

A method found to be helpful is to examine plots of the overall mean function and the functions obtained by adding and subtracting a suitable multiple of the principal component function in question.

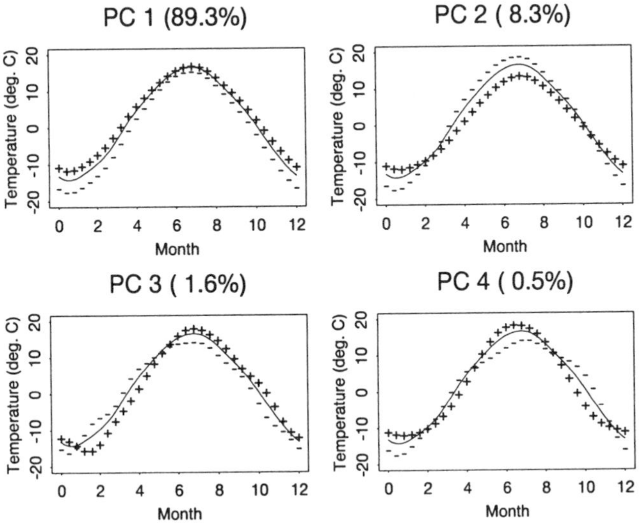

FIGURE 6.2. The mean temperature curves and the effects of adding (+) and subtracting (−) a suitable multiple of each PC curve.

Figure 6.2 shows such a plot for the temperature data. In each case, the solid curve is the overall mean temperature, and the dotted and dashed curves show the effects of adding and subtracting a multiple of each principal component curve. This considerably clarifies the effects of the first two components. We can now see that the third principal component corresponds to a time shift effect combined with an overall increase in temperature and in range between winter and summer. The fourth corresponds to an effect whereby the onset of spring is later and autumn ends earlier.

In constructing this plot, it is necessary to choose which multiple of the principal component function to use. Define a constant C to be the root-mean-square difference between $\hat{\mu}$ and its overall time average,

$$C^2 = T^{-1}\|\hat{\mu} - \bar{\mu}\|^2 \tag{6.10}$$

where

$$\bar{\mu} = T^{-1}\int \hat{\mu}(t)\,dt.$$

It is then appropriate to plot $\hat{\mu}$ and $\hat{\mu} \pm 0.2C\hat{\gamma}_j$, where we have chosen the constant 0.2 to give easily interpretable results. Depending on the overall behaviour of $\hat{\mu}$, it may be helpful to adjust the value 0.2 subjectively. But for ease of comparison between the various modes of variability, it is best to use the same constant for all the principal component functions plotted in any particular case.

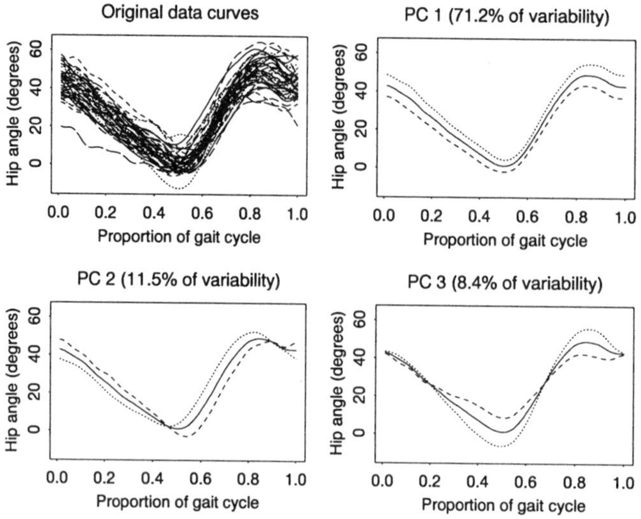

FIGURE 6.3. The hip angle observed in the gait cycles of 39 children, and the effect on the overall mean of adding and subtracting a suitable multiple of each of the first three principal component functions.

In Figure 6.3, we consider the hip angles observed during the gait of 39 children, as plotted in Figure 1.5. The angles for a single cycle are shown, along with the results of a functional PCA of these data. The effect of the first principal component of variation is approximately to add or subtract a constant to the angle throughout the gait cycle. The second component corresponds roughly to a time shift effect, which is not constant through the cycle. The third component corresponds to a variation in the overall amplitude of the angle traced out during the cycle.

6.3.2 Plotting principal component scores

An important aspect of PCA is the examination of the scores f_{im} of each curve on each component. In Figure 6.4, each weather station is identified by a four-letter abbreviation of its name given in Table 6.1. The strings are positioned roughly according to the scores on the first two principal components; some positions have been adjusted slightly to improve legibility. The West Coast stations Vancouver (VANC), Victoria (VICT) and Prince Rupert (PRUP) are in the upper right corner because they have warmer winters than most stations (high on PC 1) and less summer-winter temperature variation (high on PC 2). Resolute

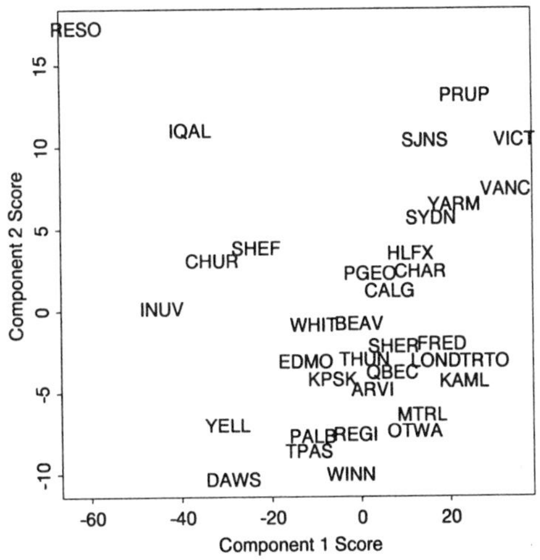

FIGURE 6.4. The scores of the weather stations on the first two principal components of temperature variation. The location of each weather station is shown by the four-letter abbreviation of its name assigned in Table 6.1.

(RESO), on the other hand, has an extremely cold winter, but does resemble the Pacific weather stations in having less summer/winter variation than some Arctic cousins, such as Inuvik (INUV).

6.3.3 Rotating principal components

In Section 6.2 we observed that the weight functions ξ_m can be viewed as defining an orthonormal set of K functions for expanding the curves to minimize a summed integrated squared error criterion (6.4). For the temperature data, for example, no set of four orthonormal functions will do a better job of approximating the curves than those displayed in Figure 6.1.

This does not mean, however, that there aren't other orthonormal sets that will do just as well. In fact, if we now use ξ to refer to the vector-valued function $(\xi_1, \dots, \xi_K)'$, then an equally good orthonormal set is defined by

$$\psi = \mathbf{T}\xi \tag{6.11}$$

where $\mathbf{T}$ is any orthonormal matrix of order K, meaning that $\mathbf{T}'\mathbf{T} = \mathbf{T}\mathbf{T}' = \mathbf{I}$. From a geometrical perspective, the vector of functions ψ is a rigid rotation of ξ. Of course, after rotation, we can no longer expect that ψ_1 will define the largest component of variation. But the point is

TABLE 6.1. The Canadian Weather Stations

Arvida, Que.	Kapuskasing, Ont.	St. John's, Nfld
Beaverlodge, B.C.	London, Ont.	Sydney, N.S.
Calgary, Alta.	Montreal, Que.	The Pas, Man.
Charlottetown, P.E.I.	Ottawa, Ont.	Thunder Bay, Ont.
Churchill, Man.	Prince Albert, Sask.	Toronto, Ont.
Dawson, Yukon	Prince George, B.C.	Vancouver, B.C.
Edmonton, Alta.	Prince Rupert, B.C.	Victoria, B.C.
Fredericton, N.B.	Quebec City, Que.	Whitehorse, Yukon
Halifax, N.S.	Regina, Sask.	Winnipeg, Man.
Inuvik, N.W.T.	Resolute, N.W.T.	Yarmouth, N.S.
Iqualuit, N.W.T.	Schefferville, Que.	Yellowknife, N.W.T.
Kamloops, B.C.	Sherbrooke, Que.	

that the orthonormal basis functions $\psi_1, \ldots, \psi_K$ are just as effective at approximating the original curves in K dimensions as their unrotated counterparts.

Can we find some rotated functions that are perhaps a little easier to interpret? Here again, we can borrow a tool that has been invaluable in multivariate analysis, VARIMAX rotation. Let $\mathbf{B}$ be a $K \times n$ matrix representing the first K principal component functions $\xi_1, \ldots, \xi_K$. For the moment, suppose that $\mathbf{B}$ has, as row m, the values $\xi_m(t_1), \ldots, \xi_m(t_n)$ for n equally spaced argument values in the interval $\mathcal{T}$. The corresponding matrix $\mathbf{A}$ of values of the rotated basis functions $\psi = \mathbf{T}\xi$ will be given by

$$\mathbf{A} = \mathbf{TB}. \tag{6.12}$$

The VARIMAX strategy for choosing the orthonormal rotation matrix $\mathbf{T}$ is to maximize the variation in the values a_{mj}^2 strung out as a single vector. Since $\mathbf{T}$ is a rotation matrix, the overall sum of these squared values will remain the same no matter what rotation we perform. In algebraic terms

$$\sum_m \sum_j a_{mj}^2 = \text{trace}\, \mathbf{A}'\mathbf{A} = \text{trace}\, \mathbf{B}'\mathbf{T}'\mathbf{T}\mathbf{B} = \text{trace}\, \mathbf{B}'\mathbf{B}.$$

Therefore, maximizing the variance of the a_{mj}^2 can happen only if these values tend either to be relatively large or relatively near zero. The values a_{mj} themselves are encouraged to be either strongly positive, near zero, or strongly negative; in-between values are suppressed. This clustering of information tends to make the components of variation easier to interpret.

There are fast and stable computational techniques for computing the rotation matrix **T** that maximizes the VARIMAX criterion. A C function for computing the VARIMAX rotation can be found through the book's world-wide web page described in Section 1.8.

Figure 6.5 displays the VARIMAX rotation of the four principal components for the temperature data. There, $n = 12$ equally spaced time points t_j were used, and the variance of the squared values $\psi_m^2(t_j)$ was maximized with respect to **T**. The resulting rotated functions ψ_m, along with the percentages of variances that they account for, are now quite different.

Collectively, the rotated functions ψ_m still account for a total of 99.7% of the variation, but they divide this variation in different proportions. The VARIMAX rotation has suppressed medium-sized values of ψ_m while preserving orthonormality. (Note that the rotated component scores are no longer uncorrelated; however, the sum of their variances is still the same, because **T** is a rotation matrix, and so they may still be considered to partition the variability in the original data.) The result is four functions that account for local variation in the winter, summer, spring and autumn, respectively. Not only are these functions much easier to interpret, but we see something new: although winter variation remains extremely important, now spring variation is clearly almost as important, about twice as important as autumn variation and over three times as important as summer variation.

Another way of using the VARIMAX idea is to let **B** contain the coefficients for the expansion of each ξ_m in terms of a basis ϕ of n functions. Thus we rotate the coefficients of the basis expansion of each ξ_m rather than rotating the values of the ξ_m themselves. Figure 6.6 shows the results using a Fourier series expansion of the principal components. The results are much more similar to the original principal components displayed in Figure 6.2. The main difference is in the first two components. The first rotated component function in Figure 6.6 is much more constant than the original first principal component, and corresponds almost entirely to a constant temperature effect throughout the year. The general shape of the second component is not changed very much, but it accounts for more of the variability, having essentially taken on part of the variability in the first unrotated component. Because the first component originally accounted for such a large proportion, 89.3%, of the variability, it is not surprising that a fairly small change in the shape of the second component results in moving about 10% of the total variability from the first to the second component. The third and fourth components are not enormously affected by the VARIMAX rotation in the Fourier domain.

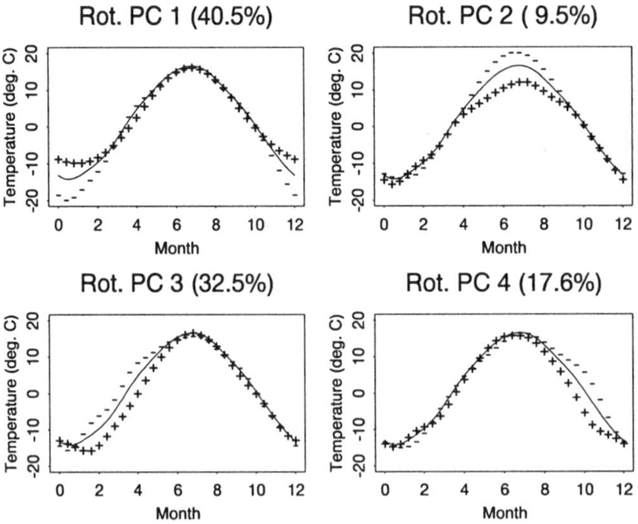

FIGURE 6.5. Weight functions rotated by applying the VARIMAX rotation criterion to weight function values, and plotted as positive and negative perturbations of the mean function.

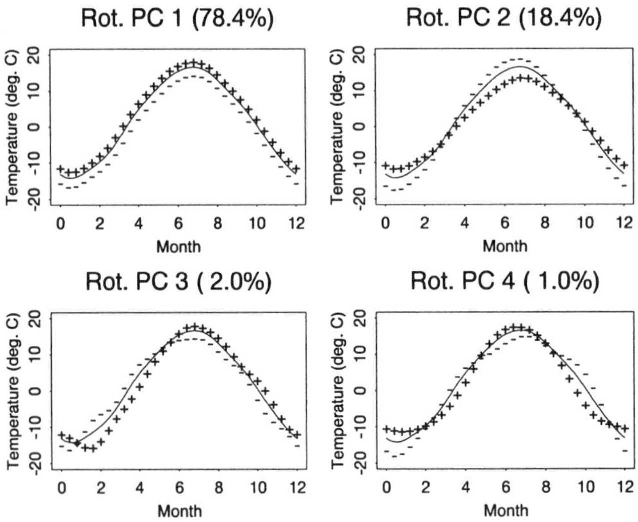

FIGURE 6.6. Weight functions rotated by applying the VARIMAX rotation criterion to weight function coefficients, and plotted as positive and negative perturbations of the mean function.

By no means is the VARIMAX criterion the only rotation criterion available. References on factor analysis and multivariate statistics such as Basilevsky (1994), Johnson and Wichern (1988), Mulaik (1972) and Seber (1984) offer a number of other possibilities. Even from the relatively brief discussion in this section, it is clear that much research remains to be done on rotation schemes tailored more directly to the functional context.

6.4 Computational methods for functional PCA

Now suppose that we have a set of N curves x_i, and that preliminary steps such as curve registration and the possible subtraction of the mean curve from each (curve centering) have been completed. Let $v(s,t)$ be the sample covariance function of the observed data. In this section, we consider possible strategies for approaching the eigenanalysis problem in (6.7). In all cases, we convert the continuous functional eigenanalysis problem to an approximately equivalent matrix eigenanalysis task.

6.4.1 Discretizing the functions

A simple approach is to discretize the observed functions x_i to a fine grid of n equally spaced values s_j that span the interval $\mathcal{T}$. This yields an $N \times n$ data matrix $\mathbf{X}$ that can be fed into a standard multivariate principal components analysis program such as the S-PLUS routine `prcomp`. This produces eigenvalues and eigenvectors satisfying

$$\mathbf{V}u = \lambda u \tag{6.13}$$

for n-vectors u.

Notice that we may well have n much larger than N. Rather than working with the $n \times n$ matrix $\mathbf{V}$, one possible approach to finding the solutions of the eigenequation (6.13) is to work in terms of the SVD $\mathbf{UDW}'$ of $\mathbf{X}$. The variance matrix satisfies $N\mathbf{V} = \mathbf{UD}^2\mathbf{U}'$, and hence the nonzero eigenvalues of $\mathbf{V}$ are the squares of the singular values of $\mathbf{X}$, and the corresponding eigenvectors are the columns of $\mathbf{U}$. If we use a standard PCA package, these steps, or corresponding ones, will be carried out automatically in any case.

How do we transform the vector principal components back into functional terms? The sample variance-covariance matrix $\mathbf{V} = N^{-1}\mathbf{X}'\mathbf{X}$ will have elements $v(s_j, s_k)$ where $v(s,t)$ is the sample covariance function. Given any function ξ, let $\tilde{\xi}$ be the n-vector of values $\xi(s_j)$. Let $w = T/n$ where T is the length of the interval $\mathcal{T}$. Then, for each s_j,

$$V\xi(s_j) = \int v(s_j, s)\xi(s)ds \approx w \sum v(s_j, s_k)\tilde{\xi}_k,$$

so the functional eigenequation $V\xi = \rho\xi$ has the approximate discrete form

$$w\mathbf{V}\tilde{\xi} = \rho\tilde{\xi}.$$

The solutions of this equation will correspond to those of (6.13), with eigenvalues $\rho = w\lambda$. The discrete approximation to the normalization $\int \xi(s)^2 ds = 1$ is $w\|\tilde{\xi}\|^2 = 1$, so that we set $\tilde{\xi} = w^{-1/2}u$ if u is a normalized eigenvector of $\mathbf{V}$. Finally, to obtain an approximate eigenfunction ξ from the discrete values $\tilde{\xi}$, we can use any convenient interpolation method. If the discretization values s_j are closely spaced, the choice of interpolation method will not usually have a great effect.

The discretization approach is the earliest approach to functional principal components analysis, used by Rao (1958, 1987) and Tucker (1958), who applied multivariate principal components analysis without modification to observed function values. We discuss the idea of discretizing the integral in more detail in Section 6.4.3, but first we consider an alternative approach that makes use of basis expansions.

6.4.2 *Basis function expansion of the functions*

One way of reducing the eigenequation (6.7) to discrete or matrix form is to express each function x_i as a linear combination of known basis functions ϕ_k. The number K of basis functions used depends on many considerations: How many discrete sampling points n were in the original data, whether some level of smoothing was to be imposed by using $K < n$, how efficient or powerful the basis functions are in reproducing the behaviour of the original functions, and so forth. For the monthly temperature data, for example, it would be logical to use a Fourier series basis orthonormal over the interval [0,12], with $K = 12$ the maximum possible dimension of the basis for the monthly temperature data, because only 12 sampling points are available per curve. Actually, for these data, a value of K as small as 7 would capture most of the interesting variation in the original data, but there is little point in reducing K below the value of 12.

Now suppose that each function has basis expansion

$$x_i(t) = \sum_{k=1}^{K} c_{ik}\phi_k(t). \tag{6.14}$$

We may write this more compactly by defining the vector-valued function x to have components $x_1, \ldots, x_N$, and the vector-valued function ϕ to have components $\phi_1, \ldots, \phi_K$. We may then express the simultaneous expansion of all N curves as

$$x = \mathbf{C}\phi$$

where the coefficient matrix $\mathbf{C}$ is $N \times K$. In matrix terms the variance-covariance function is

$$v(s,t) = N^{-1}\phi(s)'\mathbf{C}'\mathbf{C}\phi(t)$$

remembering that $\phi(s)'$ denotes the transpose of the vector $\phi(s)$ and has nothing to do with differentiation.

Define the order K symmetric matrix $\mathbf{W}$ to have entries

$$w_{k_1,k_2} = \int \phi_{k_1}(t)\phi_{k_2}(t)\,dt = \langle \phi_{k_1}, \phi_{k_2} \rangle$$

or, more compactly, $\mathbf{W} = \int \phi'\phi$. For some choices of bases, $\mathbf{W}$ will be readily available. For example, for the orthonormal Fourier series that we might use for the temperature data, $\mathbf{W} = \mathbf{I}$, the order K identity matrix. In other cases, we may have to resort to numerical integration to evaluate $\mathbf{W}$.

Now suppose that an eigenfunction ξ for the eigenequation (6.7) has an expansion

$$\xi(s) = \sum_{k=1}^{K} b_k \phi_k(s)$$

or, in matrix notation, $\xi(s) = \phi(s)'b$. This yields

$$\begin{aligned} \int v(s,t)\xi(t)\,dt &= \int N^{-1}\phi(s)'\mathbf{C}'\mathbf{C}\phi(t)\phi(t)'b\,dt \\ &= \phi(s)'N^{-1}\mathbf{C}'\mathbf{C}\mathbf{W}b. \end{aligned}$$

Therefore the eigenequation (6.7) can be expressed as

$$\phi(s)'N^{-1}\mathbf{C}'\mathbf{C}\mathbf{W}b = \rho\phi(s)'b.$$

Since this equation must hold for all s, this implies the purely matrix equation

$$N^{-1}\mathbf{C}'\mathbf{C}\mathbf{W}b = \rho b.$$

But note that $\|\xi\| = 1$ implies that $b'\mathbf{W}b = 1$ and, similarly, two functions ξ_1 and ξ_2 will be orthogonal if and only if the corresponding vectors of coefficients satisfy $b_1'\mathbf{W}b_2 = 0$. To get the required principal components, we define $u = \mathbf{W}^{1/2}b$, solve the equivalent symmetric eigenvalue problem

$$N^{-1}\mathbf{W}^{1/2}\mathbf{C}'\mathbf{C}\mathbf{W}^{1/2}u = \rho u$$

and compute $b = \mathbf{W}^{-1/2}u$ for each eigenvector.

Two special cases deserve particular attention. As already mentioned, if the basis is orthonormal, meaning that $\mathbf{W} = \mathbf{I}$, the functional PCA problem finally reduces to the standard multivariate PCA of the coefficient array $\mathbf{C}$, and we need only carry out the eigenanalysis of the order K symmetric array $N^{-1}\mathbf{C}'\mathbf{C}$.

As a rather different special case, particularly appropriate if the number of observed functions is not enormous, we may also view the observed functions x_i as their *own* basis expansions. In other words, there are N basis functions, and they happen to be the observed functions. This implies, of course, that $\mathbf{C} = \mathbf{I}$, and now the problem becomes one of the eigenanalysis of the symmetric matrix $N^{-1}\mathbf{W}$, which has entries

$$w_{ij} = \int x_i(t)x_j(t)\,dt = \langle x_i, x_j \rangle.$$

As a rule, these entries will have to be computed by some quadrature technique.

In every case, the maximum number of eigenfunctions that can in principle be computed by the basis function approach is K, the dimension of the basis. However, if the basis expansions have involved any approximation of the observed functions, then it is not advisable to use a basis expansion to K terms to calculate more than a fairly small proportion of K eigenfunctions.

The results of both the strategies we have discussed are illustrated in Figure 6.1, which shows the first four estimated eigenfunctions ξ_m of the centred temperature functions

$$x_i = \mathtt{Temp}_i - \frac{1}{35}\sum_j \mathtt{Temp}_j.$$

The smooth curves give the estimated eigenfunctions using the complete 12-term Fourier series expansion. For comparison purposes, the results of applying the discretization approach to the data are also displayed as points indicating the values of the eigenvectors. There is little discrepancy between the two sets of results. The proportions of variances for the basis function analysis turn out to be identical to those computed for the discretization approach. No attempt has been made to interpolate the discretized values to give continuous eigenfunctions, but if the Fourier series interpolation method were used, the results would be identical to the results obtained by the basis method; this is a consequence of special properties of Fourier series.

6.4.3 *More general numerical quadrature*

The eigenequation (6.7) involves the integral $\int x_i(s)\xi(s)\,ds$, and the discretization strategy is to approximate this integral by a sum of discrete values. Most schemes for numerical integration or quadrature (Stoer and Bulirsch, 1980, is a good reference) involve an approximation of the form

$$\int f(s)\,ds \approx \sum_{j=1}^{n} w_j f(s_j) \tag{6.15}$$

and the method set out in Section 6.4.1 is a fairly crude special case. We restrict our attention to linear quadrature schemes of the form (6.15). There are three aspects of the approximation that can be manipulated to meet various objectives:

- n, the number of discrete argument values s_j
- s_j, the argument values, called *quadrature points*
- w_j, the weights, called *quadrature weights*, attached to each function value in the sum.

A simple example is the *trapezoidal rule*, in which the interval of integration is divided into $n-1$ equal intervals, each of width h. The s_j are the boundaries of the interval with s_1 and s_n the lower and upper limits of integration, respectively, and the approximation is

$$\int f(s)\,ds \approx h[f(s_1)/2 + \sum_{j=2}^{n-1} f(s_j) + f(s_n)/2]. \tag{6.16}$$

Note that the weights w_j are $h/2, h, \ldots, h, h/2$ and that accuracy is controlled simply by the choice of n. The trapezoidal rule has some important advantages: the original raw data are often collected for equally spaced argument values, the weights are trivial, and although the accuracy of the method is modest relative to other more sophisticated schemes, it is often entirely sufficient for the objectives at hand. The method we set out in Section 6.4.1 is similar to the trapezoidal rule, and indeed if we use periodic boundary conditions, the methods are the same, since the values $f(s_n)$ and $f(s_1)$ are identical.

Other techniques, Gaussian quadrature schemes for example, define quadrature weights and points that yield much higher accuracy for fixed n under suitable additional conditions on the integrand. Another class of procedures chooses the quadrature points adaptively to provide more resolution in regions of high integrand curvature; for these to be relevant to the present discussion, we must choose the quadrature points once for all the functions considered in the analysis.

Applying quadrature schemes of the type (6.15) to the operator V in (6.9), yields the discrete approximation

$$V\xi \approx \mathbf{VW}\tilde{\xi} \tag{6.17}$$

where, as in Section 6.4.1, the matrix $\mathbf{V}$ contains the values $v(s_j, s_k)$ of the covariance function at the quadrature points, and $\tilde{\xi}$ is an order n vector containing values $\xi(s_j)$. The matrix $\mathbf{W}$ is a diagonal matrix with diagonal values being the quadrature weights w_j.

The approximately equivalent matrix eigenanalysis problem is then

$$\mathbf{VW}\tilde{\xi} = \rho\tilde{\xi}$$

where the orthonormality requirement is now

$$\tilde{\xi}_m'\mathbf{W}\tilde{\xi}_m = 1 \text{ and } \tilde{\xi}_{m_1}'\mathbf{W}\tilde{\xi}_{m_2} = 0, \ m_1 \neq m_2.$$

Since most quadrature schemes use positive weights, we can put the approximate eigenequation in more standard form, analogous to the calculations carried out in Section 6.4.2:

$$\mathbf{W}^{1/2}\mathbf{VW}^{1/2}u = \rho u$$

where $u = \mathbf{W}^{1/2}\tilde{\xi}$ and $u'u = 1$. Then the whole procedure is as follows:

1. Choose n, the w_j's, and the s_j's.
2. Compute the eigenvalues ρ_m and eigenvectors u_m of $\mathbf{W}^{1/2}\mathbf{VW}^{1/2}$.
3. Compute
$$\tilde{\xi}_m = \mathbf{W}^{-1/2}u_m.$$
4. If needed, use an interpolation technique to convert each vector $\tilde{\xi}_m$ to a function ξ_m.

If the number n of quadrature points is less than the number of curves N, we cannot recover more than n approximate eigenfunctions. However, many applications of PCA require only a small number of the leading eigenfunctions, and any reasonably large n will serve.

To illustrate the application of this discretizing approach, we analyse the acceleration in human growth described in Chapter 1. Each

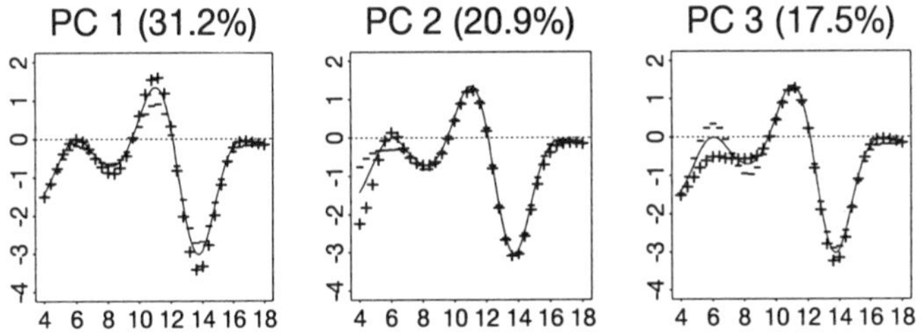

FIGURE 6.7. The solid curve in each panel is the mean acceleration in height in cm/year^2 for girls in the Zurich growth study. Each principal component is plotted in terms of its effect when added (+) and subtracted (−) from the mean curve.

curve consists of 141 equally spaced values of acceleration in height estimated for ages from 14 to 18 years, after spline smoothing and registration by certain marker events. Full details of this process can be found in Ramsay, Bock and Gasser (1995). The curves are for 112 girls who took part in the Zurich growth study (Falkner, 1960).

Figure 6.7 shows the first three eigenfunctions or harmonics plotted as perturbations of the mean function. Essentially, the first principal component reflects a general variation in the amplitude of the variation in acceleration that is spread across the entire curve, but is particularly marked during the pubertal growth spurt lasting from 10 to 16 years of age. The second component indicates variation in the size of acceleration only from ages 4 to 6, and the third component, of great interest to growth researchers, shows a variation in intensity of acceleration in the prepubertal period around ages 5 to 9 years.

6.5 Bivariate and multivariate PCA

We often wish to study the simultaneous variation of more than one function. The hip and knee angles described in Chapter 1 are an example; to understand the total system, we want to know how hip and knee angles vary jointly. Similarly, the handwriting data require the study of the simultaneous variation of the X and Y coordinates; there would be little point in studying one coordinate at a time.

6.5.1 *Defining multivariate functional PCA*

For clarity of exposition, we discuss the extension of the PCA idea to deal with bivariate functional data in the specific context of the hip and knee data. Suppose that the observed hip angle curves are $\texttt{Hip}_1, \texttt{Hip}_2, \ldots, \texttt{Hip}_n$ and the observed knee angles are $\texttt{Knee}_1, \texttt{Knee}_2, \ldots, \texttt{Knee}_n$. Let Hipmn and Kneemn be estimates of the mean functions of the Hip and Knee processes. Define v_{HH} to be the covariance operator of the $\texttt{Hip}_i$, v_{KK} that of the $\texttt{Knee}_i$, v_{HK} to be the cross-covariance function, and $v_{\mathsf{KH}}(t,s) = v_{\mathsf{HK}}(s,t)$.

A typical principal component is now defined by a 2-vector $\xi = (\xi^{\mathsf{H}}, \xi^{\mathsf{K}})'$ of weight functions, with ξ^{H} denoting the variation in the Hip curve and ξ^{K} that in the Knee curve. To proceed, we need to define an inner product on the space of vector functions. The most straightforward definition is simply to sum the inner products of the two components. We define the inner product of ξ_1 and ξ_2 to be

$$\langle \xi_1, \xi_2 \rangle = \langle \xi_1^{\mathsf{H}}, \xi_2^{\mathsf{H}} \rangle + \langle \xi_1^{\mathsf{K}}, \xi_2^{\mathsf{K}} \rangle. \tag{6.18}$$

The corresponding squared norm $\|\xi\|^2$ is simply the sum of the squared norms of the two component functions ξ^{H} and ξ^{K}.

What all this amounts to, in effect, is stringing two (or more) functions together to form a composite function. We do the same thing with the data themselves: define $\texttt{Angles}_i = (\texttt{Hip}_i, \texttt{Knee}_i)$. The weighted linear combination (6.3) becomes

$$f_i = \langle \xi, \texttt{Angles}_i \rangle = \langle \xi^{\mathsf{H}}, \texttt{Hip}_i \rangle + \langle \xi^{\mathsf{K}}, \texttt{Knee}_i \rangle. \tag{6.19}$$

We now proceed exactly as in the univariate case, extracting solutions of the eigenequation system $V\xi = \rho\xi$, which can be written out in full detail as

$$\begin{aligned} \int v_{\mathsf{HH}}(s,t)\xi^{\mathsf{H}}(t)\,dt + \int v_{\mathsf{HK}}(s,t)\xi^{\mathsf{K}}(t)\,dt &= \rho\xi^{\mathsf{H}}(s) \\ \int v_{\mathsf{KH}}(s,t)\xi^{\mathsf{H}}(t)\,dt + \int v_{\mathsf{KK}}(s,t)\xi^{\mathsf{K}}(t)\,dt &= \rho\xi^{\mathsf{K}}(s). \end{aligned} \tag{6.20}$$

In practice, we carry out this calculation by replacing each function $\texttt{Hip}_i$ and $\texttt{Knee}_i$ with a vector of values at a fine grid of points or coefficients in a suitable expansion. For each i these vectors are concatenated into a single long vector Z_i; the covariance matrix of the Z_i is a discretized version of the operator V as defined in (6.6). We carry out a standard principal components analysis on the vectors Z_i, and separate the resulting principal component vectors into the parts corresponding to Hip and to Knee. The analysis is completed

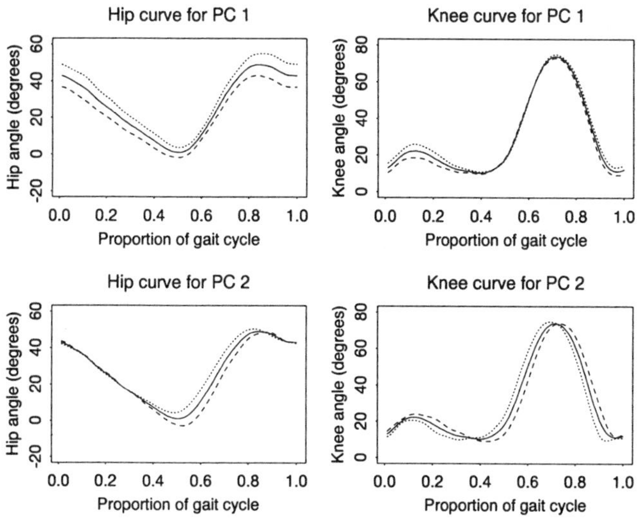

FIGURE 6.8. The mean hip and knee angle curves and the effects of adding and subtracting a multiple of each of the first two vector principal components.

by applying a suitable inverse transform to each of these parts if necessary.

If the variability in one of the sets of curves is substantially greater than that in the other, then it is advisable to consider downweighting the corresponding term in the inner product (6.18), and making the consequent changes in the remainder of the procedure. In the case of the hip and knee data, however, both sets of curves have similar amounts of variability and are measured in the same units (degrees) and so there is no need to modify the inner product.

6.5.2 *Visualizing the results*

In the bivariate case, the best way to display the result depends on the particular context. In some cases it is sufficient to consider the individual parts ξ_m^{H} and ξ_m^{K} separately. An example of this is given in Figure 6.8, which displays the first two principal components. Because $\|\xi_m^{\mathrm{H}}\|^2 + \|\xi_m^{\mathrm{K}}\|^2 = 1$ by definition, calculating $\|\xi_m^{\mathrm{H}}\|^2$ gives the proportion of the variability in the mth principal component accounted for by variation in the hip curves.

For the first principal components, this measure indicates that 85% of the variation is due to the hip curves, and this is borne out by the presentation in Figure 6.8. The effect on the hip curves of the first combined principal component of variation is virtually identical to the first principal component curve extracted from the hip curves considered alone. There is also little associated variation in the knee curves, apart from a small associated increase in the bend of the knee during the part of the cycle where all the weight is on the observed leg. The main effect of the first principal component remains an overall shift in the hip angle. This could be caused by an overall difference in stance; some people stand up more straight than others and therefore hold their trunks at a different angle from the legs through the gait cycle. Alternatively, there may simply be variation in the angle of the marker placed on the trunk.

For the second principal component, the contributions of both hip and knee are important, with somewhat more of the variability (65%) due to the knee than to the hip. We see that this principal component is mainly a distortion in the timing of the cycle, again correlated with the way in which the initial slight bend of the knee takes place. There is some similarity to the second principal component found for the hip alone, but this time there is very substantial interaction between the two joints.

A particularly effective method for displaying principal components in the bivariate case is to construct plots of one variable against the other. Suppose we are interested in displaying the mth principal component function. For equally spaced points t in the time interval on which the observations are taken, we indicate the position of the mean function values $(\mathtt{Hipmn}(t), \mathtt{Kneemn}(t))$ by a dot in the (x, y) plane, and we join this dot by an arrow to the point $(\mathtt{Hipmn}(t) + C\xi_m^{\mathsf{H}}(t),$ $\mathtt{Kneemn}(t) + C\xi_m^{\mathsf{K}}(t))$. We choose the constant C to give clarity. Of course, the sign of the principal component functions, and hence the sense of the arrows, is arbitrary, and plots with all the arrows reversed convey the same information.

This technique is displayed in Figure 6.9. The plot of the mean cycle alone demonstrates the overall shape of the gait cycle in the hip-knee plane. The portion of the plot between time points 11 and 19 (roughly the part where the foot is off the ground) is approximately half an ellipse with axes inclined to the coordinate axes. The points on the ellipse are roughly at equal angular coordinates — somewhat closer together near the more highly curved part of the ellipse. This demonstrates that in this part of the cycle, the joints are moving roughly in simple harmonic motion but with different phases. During the other part of the cycle, the hip angle is changing at a approximately constant rate as the body moves forward with the leg approximately straight, and the knee bends slightly in the middle.

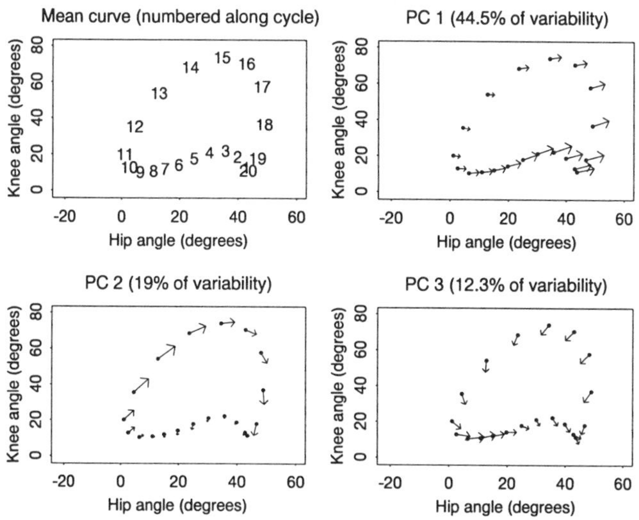

FIGURE 6.9. A plot of 20 equally spaced points in the average gait cycle, and the effects of adding a multiple of each of the first three principal component cycles in turn.

Now consider the effect of the first principal component of variation. As we have already seen, this has little effect on the knee angle, and all the arrows are approximately in the x-direction. The increase in the hip angle due to this mode of variation is somewhat larger when the angle itself is larger. This indicates that the effect contains an exaggeration (or diminution) in the amount by which the hip joint is bent during the cycle, and is also related to the overall angle between the trunk and the legs.

The second principal component demonstrates an interesting effect. There is little change during the first half of the cycle. However, during the second half, individuals with high values of this principal component would traverse roughly the same cycle but at a roughly constant time ahead. Thus this component represents a uniform time shift during the part of the cycle when the foot is off the ground.

A high score on the third component indicates two effects. There is some time distortion in the first half of the cycle, and then a shrinking of the overall cycle; an individual with a high score would move slowly through the first part of the cycle, and then perform simple harmonic motion of knee and hip joints with somewhat less than average amplitude.

7

Regularized principal components analysis

7.1 Introduction

In this chapter, we discuss the application of smoothing to functional principal components analysis. In Chapter 3 we have already seen that smoothing methods are useful in functional data analysis in preprocessing the data to obtain functional observations. The emphasis in this chapter is somewhat different, in that we incorporate the smoothing into the principal components analysis itself.

Our discussion provides a further insight into the way the method of regularization, discussed in Chapter 4, can be used rather generally in functional data analysis. As in Section 5.4.2, the basic idea is to put into practice, in any particular context, the philosophy of combining a measure of goodness-of-fit with a roughness penalty.

Consideration of the third component in Figure 6.1 indicates that some smoothing may be appropriate when estimating functional principal components. A more striking example is provided by the pinch force data discussed in Section 1.4.2. Rather than smoothing the data initially, consider the data in Figure 7.1, which consists of the original records of the force exerted by the thumb and forefinger during each of 20 brief squeezes or pinches. The observed records are not very smooth, and consequently the principal component curves in Figure 7.2 show substantial variability. There is a clear need for smoothing or

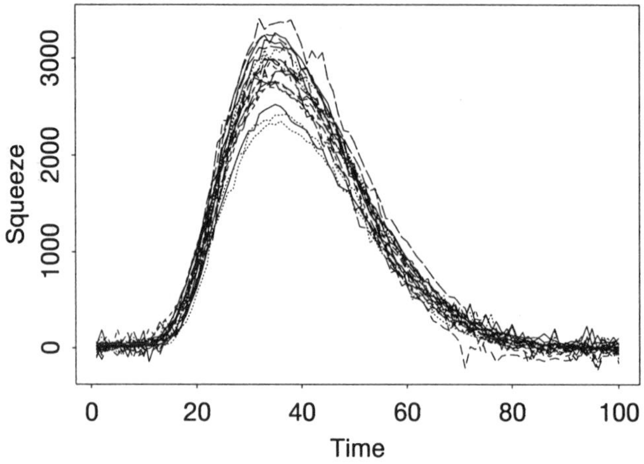

FIGURE 7.1. The aligned original recordings of the force relative to a baseline value exerted during each of 20 brief pinches.

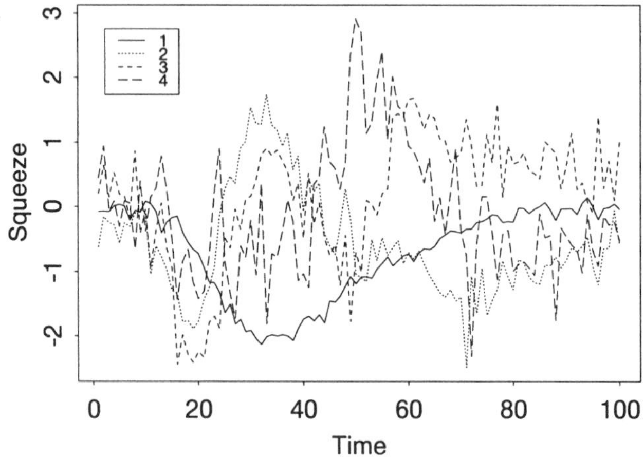

FIGURE 7.2. The first four principal component curves for the pinch force data without regularization.

regularizing of the estimated principal component curves, and we use this data set to illustrate the methodology of this chapter.

7.2 Smoothing by using roughness penalties

The main approach of this chapter is based on a roughness penalty idea, as discussed in Chapter 5. Suppose ξ is a possible principal component curve. As in standard spline smoothing, we usually penalize the roughness of ξ by its integrated squared second derivative over the interval of interest, $\text{PEN}_2(\xi) = \|D^2\xi\|^2$. We define S as the space of functions ξ for which $\|D^2\xi\|^2$ makes sense, in other words those which have continuous derivative and square-integrable second derivative, and, in addition, satisfy any relevant boundary conditions for the problem in hand, of which two examples are: (1) the periodicity of ξ and $D\xi$, and (2) $\xi(0) = D\xi(0) = 0$.

It is often useful to be able to express the roughness penalty in an alternative form. To do this, we need to impose some additional regularity conditions on the functions we are considering.

Suppose that ξ has square-integrable derivatives up to degree four, and also that ξ satisfies one of the following two conditions:

- either $D^2\xi$ and $D^3\xi$ are zero at the ends of the interval $\mathcal{T}$
- or the second and third derivatives of ξ satisfy periodic boundary conditions on $\mathcal{T}$ because we require that functions in S be periodic.

We call functions satisfying either of these conditions *very smooth* on $\mathcal{T}$. The conditions are called *natural* and *periodic* boundary conditions, respectively.

Suppose that x is any function in S and that ξ is very smooth. For notational simplicity, assume that $\mathcal{T}$ is the interval $\mathcal{T} = [0, T]$. Integrate by parts twice to obtain

$$\begin{aligned} \langle D^2x, D^2\xi\rangle &= \int_{\mathcal{T}} D^2x(s)D^2\xi(s)\,ds \\ &= Dx(T)D^2\xi(T) - Dx(0)D^2\xi(0) \\ &\quad - \int_{\mathcal{T}} Dx(s)D^3\xi(s)\,ds. \end{aligned} \tag{7.1}$$

Under natural boundary conditions, $D^2\xi(T) = D^2\xi(0) = 0$, so the boundary terms in (7.1) are both zero; under periodic boundary conditions, both Dx and $D^2\xi$ are periodic on $[0, 1]$, implying that

$$Dx(T)D^2\xi(T) = Dx(0)D^2\xi(0)$$

and the boundary terms cancel out. In either case, integrating by parts and using the boundary conditions a second time, we obtain

$$\begin{aligned}\langle D^2x, D^2\xi\rangle &= -\int_{\mathcal{T}} Dx(s)D^3\xi(s)\,ds\\ &= \{x(T)D^3\xi(T) - x(0)D^3\xi(0)\} + \int_{\mathcal{T}} x(s)D^4\xi(s)\,ds\\ &= \langle x, D^4\xi\rangle. \qquad (7.2)\end{aligned}$$

This immediately implies that the roughness penalty can be rewritten as

$$\|D^2\xi\|^2 = \langle \xi, D^4\xi\rangle \qquad (7.3)$$

which will be a useful expression in our subsequent discussion. We shall assume throughout that the PCA functions ξ we are estimating are very smooth.

7.3 Estimating smoothed principal components

Now consider the estimation of the leading principal component. In an unsmoothed functional PCA as described in Chapter 6, we would estimate this by maximizing $\langle \xi, V\xi\rangle$ subject to $\|\xi\|^2 = 1$. Here, $\langle \xi, V\xi\rangle$ is the sample variance of the principal component scores $\langle \xi, x_i\rangle$ over the observations x_i. As explained in Section 6.2.4, this maximization problem is solved by finding the leading solution of the eigenfunction equation $V\xi = \rho\xi$.

However, it may well be the case that maximizing this sample variance is not our only aim. If we are incorporating smoothing into our procedure, then we have another aim, to prevent the roughness of the estimated principal component ξ from being too large. The key to the roughness penalty approach is to make explicit this possible conflict. In the context of estimating a principal component function, there is a tradeoff between maximizing the sample variance $\langle \xi, V\xi\rangle$ and keeping the roughness penalty $\mathsf{PEN}_2(\xi)$ from becoming too large. As usual in the roughness penalty method, the tradeoff is controlled by a smoothing parameter $\lambda \geq 0$ which regulates the importance of the roughness penalty term.

Given any possible principal component function ξ with $\|\xi\|^2 = 1$, one way of penalizing the sample variance $\langle \xi, V\xi\rangle$ is to divide it by $\{1 + \lambda \times \mathsf{PEN}_2(\xi)\}$. This gives the *penalized sample variance*

$$\mathsf{PCAPSV} = \frac{\langle \xi, V\xi\rangle}{\|\xi\|^2 + \lambda \times \mathsf{PEN}_2(\xi)}. \qquad (7.4)$$

Increasing the roughness of ξ while maintaining λ fixed decreases PCAPSV, as defined in (7.4), since $\text{PEN}_2(\xi)$ increases. Moreover, PCAPSV reverts to the raw sample variance as $\lambda \to 0$. On the other hand, the larger the value of λ, the more the penalized sample variance is affected by the roughness of ξ. In the limit $\lambda \to \infty$, the component ξ is forced to be of the form $\xi = a$ in the periodic case and $\xi = a + bt$ in the nonperiodic case, for some constants a and b.

How is PCAPSV actually maximized? Making use of the relationship (7.3), we have

$$\text{PCAPSV} = \frac{\langle \xi, V\xi \rangle}{\langle \xi, (I + \lambda D^4)\xi \rangle}.$$

This expresses the penalized sample variance in the form considered in Section A.3.3. By the results described there, the maximum of PCAPSV is the eigenfunction ξ in the leading solution of the eigenfunction equation

$$V\xi = \rho(I + \lambda D^4)\xi. \tag{7.5}$$

Of course, it is usually of interest not merely to estimate the leading principal component, but also to estimate the other components. A good way of estimating these is to find *all* the eigenvalues ρ and eigenfunctions ξ of the equation (7.5). Suppose that ξ_j is the eigenfunction corresponding to the jth largest eigenvalue. Since the ξ_j's will be used as estimates of principal component functions, we rescale them to satisfy $\|\xi_j\|^2 = 1$.

How can this procedure be interpreted? As noted in Section A.3.3, the ξ_j maximizes the penalized variance (7.4) subject to $\|\xi\|^2 = 1$ and a modified form of orthogonality to the previously estimated components

$$\langle \xi_j, \xi_k \rangle + \lambda \langle D^2\xi_j, D^2\xi_k \rangle = 0 \text{ for } k = 1, \ldots, j - 1. \tag{7.6}$$

The use of the modified orthogonality condition (7.6) means that we can find the estimates of all the required eigenfunctions by solving the *single* generalized eigenproblem (7.5). Silverman (1996) provides a detailed theoretical discussion of the advantages of this approach, and a practical algorithm is set out in the next section.

7.4 Finding the regularized PCA in practice

In practice, the eigenproblem (7.5) is most easily solved by working in terms of a suitable basis. First of all, consider the periodic case, for which it is easy to set out an algorithm based on Fourier series.

7.4.1 The periodic case

Suppose, then, that $\mathcal{T}$ is the interval $[0, 1]$ and that periodic boundary conditions are valid for all the functions we are considering. In particular, this means that the data $x_i(s)$ themselves are periodic. Let $\{\phi_\nu\}$ be the series of Fourier functions defined in (3.10). For each j, define $\omega_{2j-1} = \omega_{2j} = 2\pi j$. Given any periodic function x, we can expand x as a Fourier series with coefficients $c_\nu = \langle x, \phi_\nu \rangle$, so that

$$x(s) = \sum_\nu c_\nu \phi_\nu(s) = c'\phi(s).$$

The operator D^2 has the useful property that, for each ν,

$$D^2\phi_\nu = -\omega_\nu^2 \phi_\nu,$$

meaning that we can also expand D^2x as

$$D^2x(s) = -\sum_\nu \omega_\nu^2 c_\nu \phi_\nu(s).$$

Since the ϕ_ν are orthonormal, it follows that the roughness penalty $\|D^2x\|^2$ can be written as a weighted sum of squares of the coefficients c_ν:

$$\|D^2x\|^2 = \langle -\sum_\nu \omega_\nu^2 c_\nu \phi_\nu, -\sum_\nu \omega_\nu^2 c_\nu \phi_\nu \rangle = \sum_\nu \omega_\nu^4 c_\nu^2.$$

Now proceed by expanding the data functions to sufficient terms in the basis to approximate them closely. We can use a fast Fourier transform on a finely discretized version of the observed data functions do this efficiently. Denote by c_i the vector of Fourier coefficients of the observation $x_i(s)$, so that $x_i(s) = c_i'\phi(s)$ where ϕ is the vector of basis functions. Let $\mathbf{V}$ be the covariance matrix of the vectors c_i, and let $\mathbf{S}$ be the diagonal matrix with entries

$$S_{\nu\nu} = (1 + \lambda\omega_\nu^4)^{-1/2}.$$

The matrix $\mathbf{S}$ then corresponds to a smoothing operator S.

Let y be the vector of coefficients of any principal component curve ξ, or

$$\xi(s) = \sum_\nu y_\nu \phi_\nu(s) = y'\phi(s). \tag{7.7}$$

In terms of Fourier coefficients, the equation (7.5) can be written

$$\mathbf{V}y = \rho\mathbf{S}^{-2}y \tag{7.8}$$

which can be rewritten

$$(\mathbf{SVS})(\mathbf{S}^{-1}y) = \rho(\mathbf{S}^{-1}y). \tag{7.9}$$

The matrix **SVS** is the covariance matrix of the vectors $\mathbf{S}c_i$, the Fourier coefficient vectors of the original data smoothed by the application of the smoothing operator S.

To find the solutions of (7.9), suppose that u is an eigenvector of **SVS** with eigenvalue ρ. Finding the eigenvectors and eigenvalues of **SVS** corresponds precisely to carrying out an unsmoothed PCA of the *smoothed* data $\mathbf{S}c_i$. Then it is apparent that any multiple of $\mathbf{S}u$ is a solution of (7.9) for the same ρ. Because we require $\|y\|^2 = 1$, renormalize and set $y = \mathbf{S}u/\|\mathbf{S}u\|$. The functional principal component ξ corresponding to y is then computed from (7.7).

Putting these steps together gives the following procedure for carrying out the smoothed principal component analysis of the original data:

1. Compute the coefficients c_i for the expansion of each sample function x_i in terms of basis ϕ.
2. Operate on these coefficients by the smoothing operator S.
3. Carry out a standard PCA on the resulting smoothed coefficient vectors $\mathbf{S}c_i$.
4. Apply the smoothing operator S to the resulting eigenvectors u, and renormalize so that the resulting vectors y have unit norm.
5. Compute the principal component function ξ from (7.7).

7.4.2 *The nonperiodic case*

Now turn to the nonperiodic case, where Fourier expansions are no longer appropriate because of the boundary conditions. Suppose that $\{\phi_\nu\}$ is a suitable basis for the space of smooth functions S on $[0,1]$. Possible bases include B-splines on a fine mesh, or possibly orthogonal polynomials up to some degree. In either case, we choose the dimensionality of the basis to represent the functions $x_i(s)$ well. As in the discussion of the periodic case, let c_i be the vector of coefficients of the data function $x_i(s)$ in the basis $\{\phi_\nu\}$. Let **V** be the covariance matrix of the vectors c_i.

Define **J** to be the matrix $\int \phi\phi'$, whose elements are $\langle \phi_j, \phi_k\rangle$ and **K** the matrix whose elements are $\langle D^2\phi_j, D^2\phi_k\rangle$. The penalized sample variance can be written as

$$\text{PCAPSV} = \frac{\langle \xi, V\xi\rangle}{\|\xi\|^2 + \lambda\|D^2\xi\|^2} = \frac{y'\mathbf{V}y}{y'\mathbf{J}y + \lambda y'\mathbf{K}y} \tag{7.10}$$

and the eigenequation corresponding to (7.5) is given by

$$\mathbf{V}y = \rho(\mathbf{J} + \lambda\mathbf{K})y. \tag{7.11}$$

Now perform a factorization $\mathbf{LL}' = \mathbf{J} + \lambda\mathbf{K}$ and define $\mathbf{S} = \mathbf{L}^{-1}$. We can find a suitable matrix $\mathbf{L}$ by an SVD or by the standard numerical linear algebra technique called Cholesky factorization, in which case $\mathbf{L}$ is a lower triangular matrix. The equation (7.11) can now be written as

$$(\mathbf{SVS}')(\mathbf{L}'y) = \rho\mathbf{L}'y.$$

We can now work through the same stages as for the periodic case, keeping careful track whether to use the matrix $\mathbf{S}$ or $\mathbf{S}'$ at each stage. The algorithm obtained is as follows:

1. Expand the observed data x_i with respect to the basis ϕ to obtain coefficient vectors c_i.
2. Solve $\mathbf{L}d_i = c_i$ for each i to find the vectors $\mathbf{S}c_i = d_i$.
3. Carry out a standard PCA on the coefficient vectors d_i.
4. Apply the smoothing operator $\mathbf{S}'$ to the resulting eigenvectors u by solving $\mathbf{L}'y = u$ in each case, and renormalize so that the resulting vectors y have $y'\mathbf{J}y = 1$.
5. Transform back to find the principal component functions ξ using (7.7).

If we use a B-spline basis and define $\mathbf{L}$ by a Cholesky factorization, then the matrices $\mathbf{J}$, $\mathbf{K}$ and $\mathbf{L}$ are all band matrices, and by using appropriate linear algebra routines, we can carry out all the calculations extremely economically. Even in the full matrix case, especially if not too many basis functions are used, the computations are reasonably fast because $\mathbf{S}$ never has to be found explicitly.

7.5 Choosing the smoothing parameter by cross-validation

We now turn from algorithmic considerations to the question of how to choose the smoothing parameter λ in practice. Although it is perfectly adequate for many purposes to choose the smoothing parameter subjectively, we can also use a cross-validation approach to choose the amount of smoothing automatically. Some general remarks about the use of automatic methods for choosing smoothing parameters are found in Section 3.1 of Green and Silverman (1994).

To consider how a cross-validation score could be calculated, suppose that x is an observation from the population. Then, by the optimal basis property discussed in Section 6.2.3, the principal components have the property that, for each m, an expansion in terms of the functions $\xi_1, \ldots, \xi_m$ can explain more of the variation in x than any other collection of m functions.

Now let $\mathbf{G}$ be the $m \times m$ matrix whose (i, j) element is the inner product $\langle \xi_i, \xi_j \rangle$. Then the component of x orthogonal to the subspace spanned by $\xi_1, \ldots, \xi_m$ can be expressed as

$$\zeta_m = x - \sum_{i=1}^{m} \sum_{j=1}^{m} (\mathbf{G}^{-1})_{ij} \langle \xi_i, x \rangle \xi_j.$$

If we wish to consider the efficacy of the first m components, then a measure to consider is $E\|\zeta_m\|^2$; in order not to be tied to a particular m, we can, for example, minimize $\sum_m E\|\zeta_m\|^2$. In both cases, we do not have new observations x to work with, and the usual cross-validation paradigm has to be used, as follows:

1. Subtract the overall mean from the observed data x_i.
2. For a given smoothing parameter λ, let $\xi_j^{[i]}(\lambda)$ be the estimate of ξ_j obtained from all the data except x_i.
3. Define $\zeta_m^{[i]}(\lambda)$ to be the component of x_i orthogonal to the subspace spanned by $\{\xi_j^{[i]}(\lambda) : j = 1, \ldots, m\}$.
4. Combine the $\zeta_m^{[i]}(\lambda)$ to obtain the cross-validation scores

$$\mathrm{CV}_m(\lambda) = \sum_{i=1}^{n} \|\zeta_m^{[i]}(\lambda)\|^2 \tag{7.12}$$

   and hence

$$\mathrm{CV}(\lambda) = \sum_{m=1}^{\infty} \mathrm{CV}_m(\lambda). \tag{7.13}$$

   In practice, we would of course truncate the sum in (7.13) at some convenient point. Indeed, given n data curves, we can estimate at most $n - 1$ principal components, and so the sum must be truncated at $m = n - 1$ if not at a smaller value.

5. Minimize $\mathrm{CV}(\lambda)$ to provide the choice of smoothing parameter.

Clearly there are other possible ways of combining the $\mathrm{CV}_m(\lambda)$ to produce a cross-validation score to account for more than one value of m, but we restrict attention to $\mathrm{CV}(\lambda)$ as defined in (7.13).

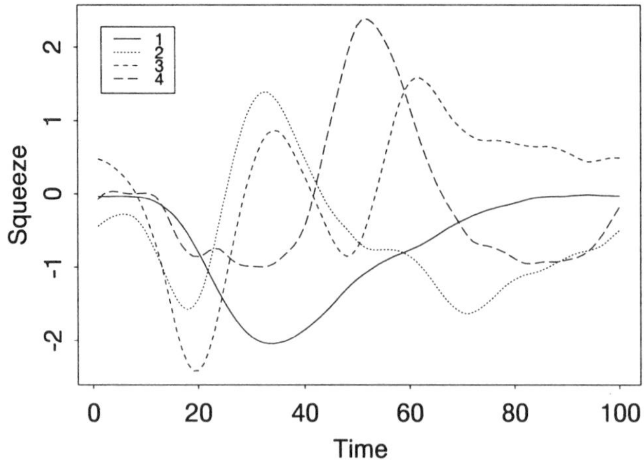

FIGURE 7.3. The first four smoothed principal components for the pinch force data, smoothed by the method of Section 7.3. The smoothing parameter is chosen by cross-validation.

7.6 An example: The pinch force data

We can now apply the smoothing method to the unsmoothed pinch force data. Figure 7.3 shows the effect of applying principal components analysis using the method for smoothed PCA set out above. We choose the smoothing parameter to minimize the score $\text{CV}(\lambda)$ defined in Section 7.5. It was found satisfactory to calculate the cross-validation score on a grid (on a logarithmic scale) of values of the smoothing parameter λ and pick out the minimum. The grid can be quite coarse, since small changes in the numerical value of λ do not make very much difference to the smoothed principal components. For this example, we calculated the cross-validation scores for $\lambda = 0$ and $\lambda = 1.5^{i-1}$ for $i = 1, \ldots, 30$, and we attained the minimum of $\text{CV}(\lambda)$ by setting $\lambda = 37$. The smoothing method achieves the aim of removing the considerable roughness in the raw principal component curves in Figure 7.2.

Figure 7.4 displays the effects on the mean curve of adding and subtracting a multiple of each of the first four smoothed principal components. The first component corresponds to an effect whereby the shape of the impulse is not substantially changed, but its overall scale is increased. The second component (with appropriate sign) corresponds

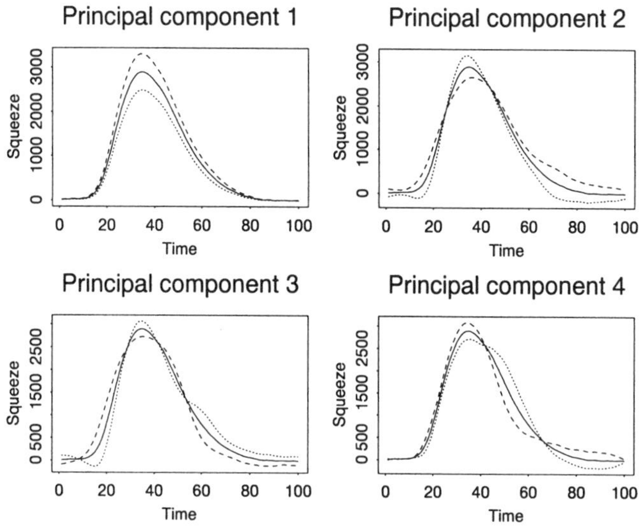

FIGURE 7.4. The effect on the overall mean curve of adding and subtracting a suitable multiple of each of the first four smoothed principal component curves provided in Figure 7.3.

roughly to a compression in the overall time scale during which the squeeze takes place. Both of these effects were removed in the analysis of Ramsay, Wang and Flanagan (1995) before any detailed analysis was carried out. It is, however, interesting to note that they occur as separate components and therefore are essentially uncorrelated with one another, and with the effects found subsequently. The third component corresponds to an effect whereby the main part takes place more quickly but the tail after the main part is extended to the right. The fourth component corresponds to a higher peak correlated with a tail-off that is faster initially but subsequently slower than the mean. The first and second effects are transparent in their interest, and the third and fourth are of biomechanical interest in indicating how the system compensates for departures from the (remarkably reproducible) overall mean. The smoothing we have described makes the effects very much clearer than they are in the raw principal component plot.

The estimated variances σ^2 indicate that the four components displayed respectively explain 86.2%, 6.7%, 3.5% and 1.7% of the variability in the original data, with 1.9% accounted for by the remaining components. The individual principal component scores indicate that there is one curve with a fairly extreme value of principal component 2

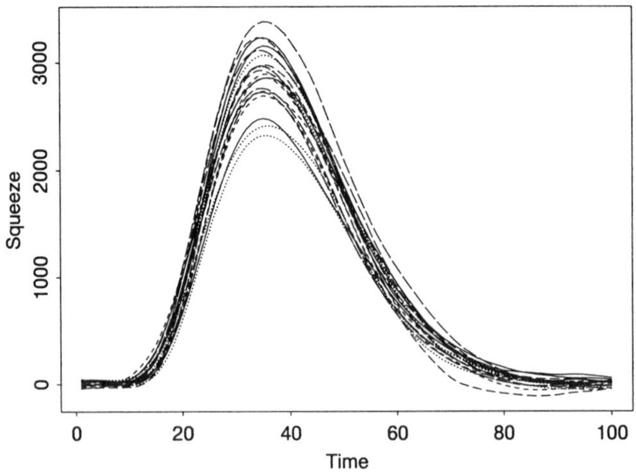

FIGURE 7.5. The pinch force data curves, smoothed by a roughness penalty method with the same smoothing parameter as used for the smoothed PCA, and with the baseline pressure subtracted.

(corresponding to moving more quickly than average through the cycle) but this curve is not unusual in other respects.

7.7 Smoothing the data rather than the PCA

In this section, we compare the method of regularized principal components analysis with an approach akin to that discussed earlier in the book. Instead of carrying out our smoothing step within the PCA, we smooth the data first, and then carry out an unsmoothed PCA. This approach to functional PCA was taken by Besse and Ramsay (1986), Ramsay and Dalzell (1991) and Besse, Cardot and Ferraty (1997). Of course, conceivably any smoothing method can be used to smooth the data, but to make a reasonable comparison, we use a roughness penalty smoother based on integrated squared second derivative. For simplicity, let us restrict our attention to the case of periodic boundary conditions.

Suppose that x is a data curve, and that we regard x as the sum of a smooth curve and a noise process. We would obtain the roughness penalty estimate of the smooth curve by minimizing

$$\text{PENRSS} = \|x - g\|^2 + \lambda \|D^2 g\|^2$$

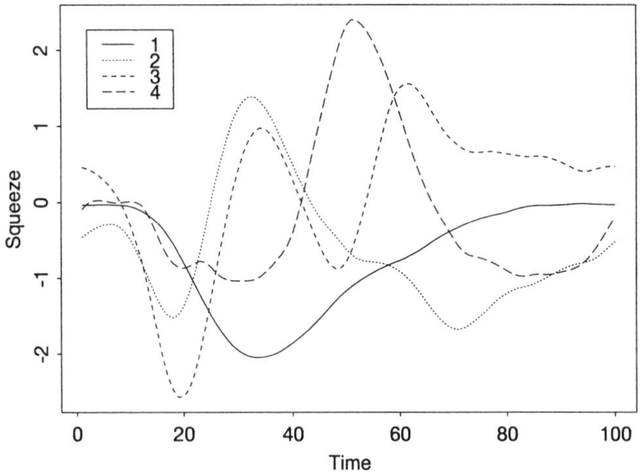

FIGURE 7.6. The first four principal component curves of the smoothed data as shown in Figure 7.5.

over g in S. As usual, λ is a smoothing parameter that controls the tradeoff between fidelity to the data and smoothing. This is a generalization of the spline smoothing method discussed in Chapter 4 to the case of functional data.

Consider an expansion of x and g in terms of Fourier series as in Section 7.4.1, and let c and d be the resulting vectors of coefficients. Then

$$\text{PENRSS} = \|c - d\|^2 + \lambda \sum_{\nu} \omega_{\nu}^4 d_{\nu}^2,$$

and hence the coefficients of the minimizing g satisfy

$$d = \mathbf{S}^2 c \tag{7.14}$$

where $\mathbf{S}$ is as defined in Section 7.4.1. Note that this demonstrates that the smoothing operator $\mathbf{S}$ used twice in the algorithm set out in Section 7.4.1 can be regarded as a half-spline-smooth, since $\mathbf{S}^2$ is the operator corresponding to classical spline smoothing.

Now let us consider the effect of smoothing the data by the operator $\mathbf{S}^2$ using the same smoothing parameter $\lambda = 37$ as in the construction of Figures 7.3 and 7.4. The effect of this smoothing on the data is illustrated in Figure 7.5. Figure 7.6 shows the first four principal component curves of the smoothed data. Although the two methods do not give identical results, the differences between them are too small to affect any interpretation.

However, this favourable comparison depends rather crucially on the way in which the data curves are smoothed, and in particular on the match between the smoothing level implied in (7.14) and the smoothing level used for the PCA itself. For example, we tried smoothing the force functions curves individually, selecting the smoothing parameters by the generalized cross-validation approach used in the S-PLUS function `smooth.spline`. The result was much less successful, in the sense that the components were far less smooth. The reason appears to be that this smoothing technique tended to choose much smaller values of the smoothing parameter λ.

Kneip (1994) considers several aspects of an approach that first smooths the data and then extracts principal components. Under a model where the data are corrupted by a white noise error process, he investigates the dependence of the quality of estimation of the principal components on both sample size and sampling rate. In an application based on economics data, he shows that smoothing is clearly beneficial in a practical sense.

7.8 The procedure of Rice and Silverman

Another approach to the smoothing of functional PCA was set out by Rice and Silverman (1991). They considered a stepwise procedure incorporating the roughness penalty in a different way. Their proposal requires a separate smoothing parameter λ_j for each principal component. The principal components are estimated successively. The estimate $\xi_j^\dagger$ of ξ_j being found by maximizing $\langle \xi, V\xi \rangle - \lambda_j \|D^2\xi\|^2$ subject to the conventional orthonormality conditions $\|\xi\|^2 = 1$ and $\langle \xi, \xi_k^\dagger \rangle = 0$ for $k = 1, \ldots, j-1$.

This approach is computationally more complicated because a separate eigenproblem has to be posed and solved for each principal component; for more details, see the original paper. Theoretical results in Pezzulli and Silverman (1993) and Silverman (1996) also suggest that the procedure described in Section 7.2 is likely to be advantageous under conditions somewhat milder than those for the Rice-Silverman procedure.

8

Principal components analysis of mixed data

8.1 Introduction

It is a characteristic of statistical methodology that problems do not always fall into neat categories. In the context of the methods discussed in this book, we often have *both* a vector of data *and* an observed function on each individual of interest. In this chapter, we consider some ways of approaching such mixed data, extending the ideas of PCA that we have already developed.

In Chapter 5 we have discussed one way in which mixed data can arise. In certain cases, it is appropriate or desirable to register the observed curves with one another, for example by shifting each one in time or by adding a constant to each one. This procedure yields mixed data: The vector of data consists of the parameters of the transformation that best maps the observed function to the overall mean, and the transformed function can be thought of as a functional observation in its own right. The methods we discuss in this chapter incorporate this registration process into the principal components analysis, and we illustrate our methodology using the Canadian temperature data as an example.

Of course, there are many other situations where we have numerical observations as well as functional observations on the individuals of interest, and the PCA methodology we set out can be easily generalized to deal with them.

8.2 General approaches to mixed data

Suppose that we observe a function x_i and a number (or a vector of numbers) y_i for each individual in our sample, How might we use PCA to analyse such data?

There are three different ways of viewing the y_i. First, it may be that the y_i are simply nuisance parameters, of no real interest to us in the analysis, for example corresponding to the time at which a recording instrument is activated. In this case we would quite simply ignore them. The y_i can be thought of as one of the features of almost all real data sets that we choose not to include in the analysis.

On the other hand, both the functions x_i and the observations y_i may be of primary importance. This is the case to which we give the most attention, from Section 8.3 onwards. Thus, we have *hybrid* data (x_i, y_i), and investigate to what extent our PCA methodology can be extended to this situation. There is some connection with the methodology described in Section 6.5 for bivariate curve data with values $(x_i(t), y_i(t))$, though the data under present consideration consist of pairs where only one component of the vector (x_i, y_i) is itself a function.

As a third and somewhat intermediate possibility, the y_i may be of marginal importance, our central interest being in the functions x_i. In this case, we could ignore the y_i initially and carry out a PCA of the curves $x_i(t)$ alone. Having done this, we could investigate the connection between the scores on the principal component scores and the variable(s) y_i. We could calculate the sample correlations between the principal component scores and the components of the y_i. Alternatively or additionally, we could plot the y_i against the principal component scores or use other methods for investigating dependence. In this general approach, the y_i would not have been used in the first part of the analysis itself; however, they would have played a key part in interpreting the analysis. It would be interesting, for example, to notice that a particular principal component of the x_i was highly correlated with y_i. We develop this approach further in Section 8.5.

8.3 The PCA of hybrid data

8.3.1 *Combining function and vector spaces*

Now suppose that our observations consist of pairs (x_i, y_i), where x_i is a function on the interval $\mathcal{T}$ and y_i is a vector of length M. How would we define a principal component of variation of such data? A typical principal component weight function would consist of a *pair* (ξ, v), where v is an M-vector, and the principal component score of a particular observation would then be the sum

$$\eta_i = \langle x_i, \xi \rangle + \langle y_i, v \rangle = \int x_i(s)\xi(s)\,ds + y_i' v. \tag{8.1}$$

Another way of saying this is that the principal component would be made up of a functional part ξ and a vector part v, corresponding to the functional and vector (or numerical) parts of the original data. A typical observation from the distribution of the data would be modelled as

$$\begin{pmatrix} x_i \\ y_i \end{pmatrix} = \sum_j \eta_{ij} \begin{pmatrix} \xi_j \\ v_j \end{pmatrix}, \tag{8.2}$$

where (ξ_j, v_j) is the jth principal component weight and, as j varies, the vectors of principal component scores $\eta_{ij} = \langle x_i, \xi_j \rangle + \langle y_i, v_j \rangle$ are uncorrelated variables with mean zero.

This kind of hybrid data PCA can very easily be dealt with in our general functional framework. Define $\mathcal{Z}$ to the space of pairs $z = (x, y)$, where x is a smooth function and y is a vector of length M. Given any two elements $z^{(1)} = (x^{(1)}, y^{(1)})$ and $z^{(2)} = (x^{(2)}, y^{(2)})$ of $\mathcal{Z}$, define the inner product

$$\langle z^{(1)}, z^{(2)} \rangle = \langle x^{(1)}, x^{(2)} \rangle + \langle y^{(1)}, y^{(2)} \rangle \tag{8.3}$$

where the inner product between the x's is the usual functional inner product $\int x^{(1)} x^{(2)}$, and the inner product of the y's is the vector inner product $y^{(1)\prime} y^{(2)}$. From (8.3) we can define the norm $\|z\|^2 = \langle z, z \rangle$ of any z in $\mathcal{Z}$.

Now that we have defined an inner product and norm on $\mathcal{Z}$, write z_i for the data pair (x_i, y_i). To find the leading principal component, we wish to find $\zeta = (\xi, v)$ in $\mathcal{Z}$ to maximize the sample variance of the $\langle \zeta, z_i \rangle$ subject to $\|\zeta\|^2 = 1$. The $\langle \zeta, z_i \rangle$ are of course exactly the same as the quantities $\eta_i = \int x_i(s)\xi(s)\,ds + y_i' v$ specified in equation (8.1).

Subsequent principal components maximize the same sample variance subject to the additional condition of orthogonality to the principal components already found, orthogonality being defined by the hybrid inner product (8.3). Principal components found in this way yield principal component scores that are uncorrelated, just as for conventional multivariate PCA.

The PCA of hybrid data is thus very easily specified in principle. However, there are several important issues raised by this idea, and we discuss these in the following sections.

8.3.2 *Finding the principal components in practice*

How do we carry out the constrained maximization of the sample variance of the $\langle \zeta, z_i \rangle$ in practice? Suppose that ϕ_k is a basis of K functions in which the functional parts x_i of the hybrid data z_i can be well approximated. Given any element $z = (x, y)$ of $\mathcal{Z}$, define the K-vector c to be the coefficients of x relative to the basis ϕ_k. Now let $p = K + M$, and let w be the p-vector

$$w = \begin{bmatrix} c \\ y \end{bmatrix}.$$

Suppose that the basis ϕ_k is an orthonormal basis, the Fourier functions, for example. Then the inner product (8.3) of any two elements $z^{(1)}$ and $z^{(2)}$ of $\mathcal{Z}$ is precisely equal to the ordinary vector inner product $\langle w^{(1)}, w^{(2)} \rangle$ of the corresponding p-vectors of coefficients. Thus, if we use this method of representing members of $\mathcal{Z}$ by vectors, we have a representation in which the vectors behave exactly as if they were p-dimensional multivariate observations, with the usual Euclidean inner product and norm. It follows that we can use standard multivariate methods to find the PCA.

In summary, we can proceed as follows to carry out a PCA:

1. For each i, let c_i be the vector of the first K Fourier coefficients of x_i.

2. Augment each c_i by y_i to form the p-vector w_i.

3. Carry out a standard PCA of the w_i, by finding the eigenvalues and eigenvectors of the matrix $N^{-1} \sum_i w_i w_i'$.

4. If u is any resulting eigenvector, the first K elements of u are the Fourier coefficients of the functional part of the principal component, and the remaining elements are the vector part.

Since the procedure we have set out is a generalization of ordinary functional PCA, we may wish to incorporate some smoothing, and this is discussed in the next section.

8.3.3 *Incorporating smoothing*

To incorporate smoothing into our procedure, we can easily generalize the smoothing methods discussed in Chapter 7. The key step in the method is to define the roughness of an element $z = (x, y)$ of $\mathcal{Z}$. Let us take the roughness of z to be that of the functional part x of z, without any reference to the vector part y. To do this, define D^2z to be equal to the element $(D^2x, 0)$ of $\mathcal{Z}$ so that the roughness of z can then be written $\|D^2z\|^2$, just as in the ordinary functional case. The norm is taken in $\mathcal{Z}$, but since the vector part of D^2z is defined to be zero, $\|D^2z\|^2 = \|D^2x\|^2$ as required.

Once we have defined the roughness of z, we can proceed to carry out a smoothed PCA using exactly the same ideas as in Chapter 7. The smoothed principal components are found by solving the eigenequation

$$V\zeta = \rho(I + \lambda D^4)\zeta$$

for ζ in $\mathcal{Z}$, and all the general ideas of Chapter 7 carry over. As far as algorithms are concerned, the Fourier transform algorithm for the periodic case requires slight modification. If $\mathbf{z}$ is the vector representation of an element z, then the first K elements of $\mathbf{z}$ refer to the functional part, and so the roughness of z is $\sum_{k=0}^{K-1} \omega_k^4 z_k^2$, rather than allowing the sum to extend over all k. It follows that the matrix $\mathbf{S}$ used in the algorithm described in Section 7.4.1 must be modified to have diagonal elements $(1 + \lambda\omega_k^4)^{-1/2}$ for $k < K$, and 1 for $K \leq k < p$.

Apart from this modification, and of course the modified procedures for mapping between the function/vector and basis representations of elements of $\mathcal{Z}$, the algorithm is exactly the same as in Section 7.4.1. Furthermore, the way in which we can apply cross-validation to choose the smoothing parameter is the same as in Section 7.5.

To deal with the nonperiodic case, we modify the algorithm of Section 7.4.2 in the same way. The matrix $\mathbf{J}$ is a block diagonal matrix where the first K rows and columns have elements $\langle \phi_j, \phi_k \rangle$ and the last M rows and columns are the identity matrix of order M. The matrix $\mathbf{K}$ has elements $\langle D^2\phi_j, D^2\phi_k \rangle$ in its first K rows and columns, and zeroes elsewhere.

8.3.4 Balance between functional and vector variation

Readers who are familiar with PCA may have noted one potential difficulty with the methodology set out above. The variations in the functional and vector parts of a hybrid observation z are really like chalk and cheese: They are measured in units which are almost inevitably not comparable, and therefore it may well not be appropriate to weight them as we have. In the registration example, the functional part consists of the difference between the pattern of temperature on the transformed time scale and its population mean; the vector part is made up of the parameters of the time transformation. Clearly, these are not measured in directly compatible units!

One way of noticing the effect of noncomparability is to consider the construction of the inner product (8.3) on $\mathcal{Z}$, which we defined by adding the inner product of the two functional parts and that of the two vector parts. In many problems, there is no intrinsic reason to give these two inner products equal weight in the sum, and a more general inner product we could consider is

$$\langle z^{(1)}, z^{(2)} \rangle = \langle x^{(1)}, x^{(2)} \rangle + C^2 \langle y^{(1)}, y^{(2)} \rangle \tag{8.4}$$

for some suitably chosen constant C. Often, the choice of C (for example $C = 1$) is somewhat arbitrary, but we can make some remarks that may guide its choice.

First, if the interval $\mathcal{T}$ is of length $|\mathcal{T}|$, then setting $C^2 = |\mathcal{T}|$ gives the same weight to overall differences between $x^{(1)}$ and $x^{(2)}$ as to differences of similar size in a single component of the vector part y. If the measurements are of cognate or comparable quantities, this may well be a good method of choosing C. On the other hand, setting $C^2 = |\mathcal{T}|/M$ tends to weight differences in functional parts the same as differences in all vector components.

Another approach, corresponding to the standard method of PCA relative to correlation matrices, is to ensure that the overall variability in the functional parts is given weight equal to that in the vector part. To do this, we would set

$$C^2 = \frac{\sum_i \|x_i - \bar{x}\|^2}{\sum_i \|y_i - \bar{y}\|^2}$$

taking the norm in the functional sense in the numerator, and in the usual vector sense in the denominator.

Finally, in specific problems, there may be a particular rationale for some other choice of constant C^2, an example of which is discussed in Section 8.4.

Whatever the choice of C^2, the most straightforward algorithmic approach is to construct the vector representation z of any element $z = (x, y)$ of $\mathcal{Z}$ to have last M elements Cy, rather than just y. The first K elements are the coefficients of the representation of x in an appropriate basis, as before. With this modification, we can use the algorithms set out above. Some care must be taken in interpreting the results, however, because any particular principal component weight function has to be combined with the data values using the inner product (8.4) to get the corresponding principal component scores.

8.4 Combining registration and PCA

8.4.1 *Expressing the observations as mixed data*

We now return to the special case of mixed data obtained by registering a set of observed curves. For the moment, concentrate on data that may be assumed to be periodic on $[0, 1]$. We suppose that an observation can be modelled as

$$x(t + \tau) = \mu(t) + \sum_j \eta_j \xi_j(t) \tag{8.5}$$

for a suitable sequence of orthonormal functions ξ_j, and where η_j are uncorrelated random variables with mean zero and variances σ_j^2. The model (8.5) differs from the usual PCA model in allowing for a shift in time τ as well as for the addition of multiples of the principal component functions. Because of the periodicity, the shifted function $x(t + \tau)$ may still be considered as a function on $[0, 1]$.

Given a data set $x_1, \ldots, x_n$, we can use the Procrustes approach set out in Chapter 5 to obtain an estimate $\hat{\mu}$ of μ and to give values of the shifts $\tau_1, \ldots, \tau_n$ appropriate to each observation. Then we can regard the data as pairs $z_i = (\tilde{x}_i, \tau_i)$, where the τ_i are the estimated values of the shift parameter and the $\tilde{x}_i$ are the shifted mean-corrected temperature curves with values $x_i(t + \tau_i) - \hat{\mu}(t)$. Recall that a consequence of the Procrustes fitting is that the $\tilde{x}_i$ satisfy the orthogonality property

$$\langle \tilde{x}_i, D\hat{\mu} \rangle = 0. \tag{8.6}$$

8.4.2 *Balancing temperature and time shift effects*

We can now consider the different approaches set out in Section 8.2 to the analysis of the mixed data z_i. We do not discuss in any detail the approach where the vector components are simply discarded, because that reduces to the case of pure functional data already considered in previous chapters. Anyway, it is clear that it would *not* be appropriate to ignore the transformation parameters in the weather example, and so we consider the approaches that take account of them in the analysis.

We consider mainly the methodology set out in Section 8.3, and seek principal components (ξ, v) that have two effects within the model (8.5): the addition of the function ξ to the overall mean $\hat{\mu}$, together with a contribution of v to the time shift τ.

In the special case of the registration data, there is a natural way of choosing the constant C^2 that controls the balance between the functional and shift components in the inner product (8.4). Suppose that x is a function in the original data function space, and that $z = (\tilde{x}, \tau)$ is the corresponding pair in $\mathcal{Z}$, so that

$$x(t) = \hat{\mu}(t - \tau) + \tilde{x}(t - \tau).$$

Because of the orthogonality property (8.6), we can confine attention to $\tilde{x}$ that are orthogonal to $\hat{\mu}$.

To define a norm on $\mathcal{Z}$, a requirement is that, at least to first order,

$$\|z\|^2 \approx \|x - \hat{\mu}\|^2 = \int [x(s) - \hat{\mu}(s)]^2 \, ds, \tag{8.7}$$

the standard squared function norm for $x - \hat{\mu}$. This means that the norm of any small perturbation of the mean function $\hat{\mu}$ must the same, whether it is specified in the usual function space setting as $x - \hat{\mu}$, or expressed as a pair z in $\mathcal{Z}$, consisting of a perturbation $\tilde{x}$ orthogonal to $\hat{\mu}$ and a time shift.

Suppose $\|\tilde{x}\|$ and τ are small. If we let

$$C^2 = \|D\hat{\mu}\|^2 \tag{8.8}$$

then, to first order in $\|\tilde{x}\|$ and τ,

$$x(t) - \hat{\mu}(t) \approx -\tau D\hat{\mu}(t - \tau) + \tilde{x}(t - \tau).$$

By the orthogonality of $\tilde{x}$ and $D\hat{\mu}$,

$$\|z - \hat{\mu}\|^2 \approx \int \tilde{x}^2(s) + C^2\tau^2(s) \, ds = \|\tilde{x}\|^2 + C^2\|\tau\|^2, \tag{8.9}$$

as required.

With this calculation in mind, we perform our PCA of the pairs $(\tilde{x}_i, \tau_i)$ relative to the inner product (8.4) with $C^2 = \|D\hat{\mu}\|^2$.

8.4.3 *Application to the weather data*

The application of this procedure to the Canadian temperature data is set out in Figure 8.1. For each PC, it illustrates the effect on the overall mean $\hat{\mu}$ of adding and subtracting a suitable multiple of the ξ_j. Note that $\hat{\mu}$ is now the mean of the *registered* temperature curves, not of the original curves. We use the same multiple of the PC curve in each case, but of course it is no longer true that the square integral of the PC curves is equal to one. In particular, the curve part of principal component 3 yields much less variability about the mean curve than the other principal components. In each case, the sign of the principal

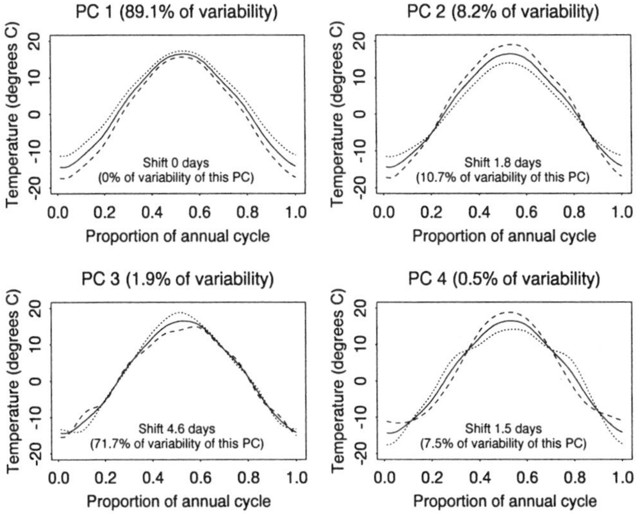

FIGURE 8.1. The mean Canadian temperature curve and the effects of adding and subtracting a suitable multiple of each PC curve, with the shift considered as a separate parameter.

component has been taken to make the shift positive; this is by no means essential, but it leads to some simplicity of interpretation.

Each panel shows the shift part of the corresponding principal component. We see that principal component 3 has a large shift component. The percentage of variability of each principal component due to the shift is worked out as $100C^2v_j^2$ with C defined as in (8.8). A principal component that is entirely shift would correspond to a shift of ± 5.4 days.

A comparison between Figure 8.1 and Figure 6.2 is instructive. First, we see that the percentage of variation explained by each of the first four principal components is very similar, but not identical, in the two analyses. This is a consequence of using the inner product derived in Section 8.4.2, which ensures approximate compatibility between the quantification of variation caused simply by the addition of a curve to the overall mean, and variation that also involves a time shift. Because the shift component has been explicitly separated out, less skill is needed to interpret the principal components. Principal component 2 shows a small shift associated with a reduction of the annual temperature range. The large shift in principal component 3 is associated with a small increase in the midsummer temperature.

8.4.4 Taking account of other effects

In the temperature example, the shift effect is not necessarily the only effect that can be extracted explicitly and dealt with separately in the functional principal components analysis. We must also take account of the overall annual average temperature for each weather station, and we do this by extending the model (8.5) to a model of the form

$$x(t+\tau) - \theta = \alpha + \mu(t) + \sum_j \eta_j \xi_j(t) \tag{8.10}$$

where θ is an annual temperature effect with zero population mean. The η_j are assumed to be uncorrelated random variables with mean zero. The parameter α is the overall average temperature (averaged both over time and over the population). For identifiability we assume that $\int \mu(s)\, ds = 0$.

The data we would use to fit such a model consist of triples $(\check{x}_i, \tau_i, \theta_i)$, where $\check{x}_i$ are the observed temperature curves registered to one another by shifts τ_i, and with each curve modified by subtracting its overall annual average $\hat{\alpha} + \theta_i$. Here the number $\hat{\alpha}$ is the time average of all the temperatures observed at all weather stations, and the individual θ_i therefore sum to zero. Because the annual average $\hat{\alpha} + \theta_i$ has been subtracted from each curve $\check{x}_i$, the curves $\check{x}_i$ each integrate to zero as well as satisfying the orthogonality condition (8.6). The mean curve $\hat{\mu}$ is then an estimate of the mean of the registered curves $\check{x}_i$, most straightforwardly the sample mean. In the hybrid data terms we have set up, the functional part of each data point is the curve $\check{x}$, whereas the vector part is the 2-vector $(\tau_i, \theta_i)'$.

To complete the specification of (8.10) as a hybrid data principal components model, we regard τ and θ as random variables which can be expanded for the same η_j, as

$$\tau = \sum_j \eta_j v_j \text{ and } \theta = \sum_j \eta_j u_j,$$

where the v_j and u_j are fixed quantities. Thus, the jth principal component is characterized by a triple (ξ_j, v_j, u_j), constituting a distortion of the mean curve by the addition of a multiple of ξ_j, together with shifts in time and in overall temperature by the same multiples of v_j and u_j, respectively.

Just as before, we carry out a PCA of the hybrid data $\{(\check{x}_i, \tau_i, \theta_i)\}$ with respect to a suitably chosen norm. To define the norm of a triple $(\check{x}, \tau, \theta)$, consider the corresponding unregistered and uncorrected curve x, defined by

$$x(t+\tau) = \hat{\alpha} + \theta + \hat{\mu}(t) + \check{x}(t).$$

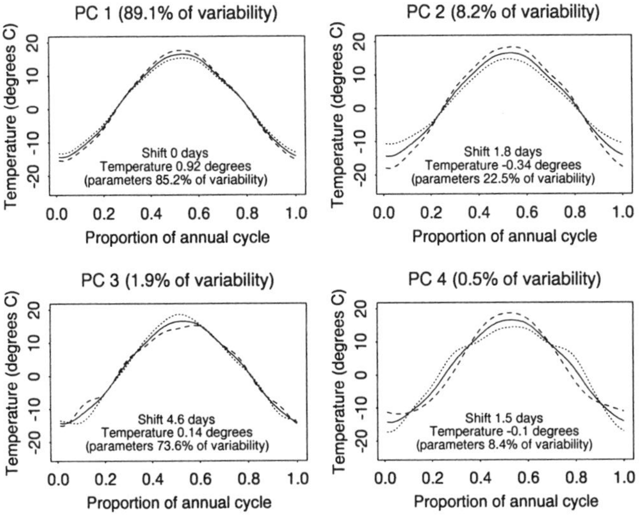

FIGURE 8.2. The mean Canadian temperature curve and the effect of adding and subtracting a suitable multiple of each PC curve, with the shift and annual average temperature considered as separate parameters.

Define $C_1 = \|D\hat{\mu}\|^2$ and $C_2 = |\mathcal{T}|$. Assume that $\breve{x}$ integrates to zero and satisfies (8.6).

By arguments similar to those used previously, using the standard square integral norm for $\breve{x}$,

$$\|x - \hat{\mu}\|^2 \approx \|\breve{x}\|^2 + C_1^2\tau^2 + C_2^2\theta^2.$$

Thus an appropriate definition of the norm of the triple is given by

$$\|(\breve{x}, \tau, \theta)\|^2 = \|\breve{x}\|^2 + C_1^2\tau^2 + C_2^2\theta^2.$$

In practice, a PCA with respect to this norm is carried out by the same general approach as before. For each i, the function $\breve{x}_i$ is represented by a vector $\breve{c}_i$ of its first K Fourier coefficients. The vector is augmented by the two values $C_1\tau_i$ and $C_2\theta_i$ to form the vector z_i. We then carry out a standard PCA on the augmented vectors z_i. The resulting principal component weight vectors are then unpacked into the parts corresponding to ξ_j, v_j and u_j, and the appropriate inverse transforms applied—just dividing by C_1 and C_2 respectively in the case of the shift and overall temperature effects, and applying an inverse Fourier transform to the first K components of the vector to find ξ_j.

Figure 8.2 shows the effect of this approach applied to the Canadian temperature data. Notice that a component that was entirely variation in overall temperature would have a temperature effect of ± 1 degree, because time is scaled to make the cycle of unit length (with time measured in years) so that $C_2 = 1$. Because each principal component is scaled to have unit norm, the maximum possible value of $(C_2 u_i)^2$ is 1, with equality if and only if the other components are zero. Similarly, since $C_1 = 365/5.4$, a component that was entirely a time shift would have $v_i = \pm 5.4/365$ years, i.e. ± 5.4 days.

In each case in the figure, the proportions of variability due to the two parametric effects, shift and overall average temperature, are combined to give the percentage of variability due to the vector parameters. Principal component 1 is almost entirely due to the variation in overall temperature, with a small effect corresponding to a decrease in range between summer and winter. (Recall that the dotted curve corresponds to a positive multiple of the principal component curve ξ_i, and the dashed curve to a negative multiple.) Principal component 2 has some shift component, a moderate negative temperature effect, and mainly comprises the effect of a decreased annual temperature range. Within this component, overall average temperature is positively associated with increased range, whereas in component 1 the association was negative. Principal component 1 accounts for a much larger proportion of the variability in the original data, and a slightly different approach in Section 8.5 shows that within the data as a whole, increased overall temperature is negatively correlated with higher range between summer and winter—colder places have more extreme temperatures.

Neither principal components 3 nor 4 contains much of an effect due to overall temperature. As before, component 3 is very largely shift, whereas component 4 corresponds to an effect unconnected to shift or overall temperature.

8.5 Separating out the vector component

This section demonstrates the other procedure suggested in Section 8.2. We carry out a principal components analysis on the *registered* curves $\check{x}_i$ and then investigate the relationship between the resulting principal component scores and the parameters τ_i and θ_i arising in the registration process. Thus we analyse only the functional part of the mixed data, and the vector part is only considered later.

The effect of doing this is demonstrated in Figure 8.3. Removing the temperature and shift effects accounts for 79.2% of the variability in the original data, and the percentages of variability explained by the various principal components have been multiplied by 0.208, to make

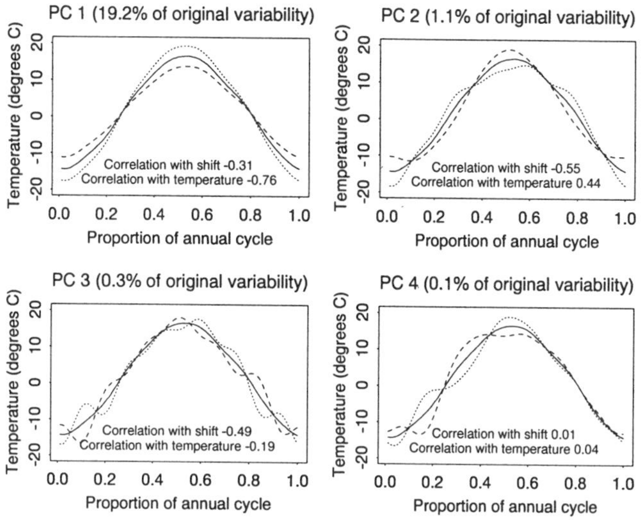

FIGURE 8.3. Principal component analysis carried out on the Canadian temperature curves adjusted for time shift and for annual average temperature.

them express parts of the variability of the original data, rather than the adjusted data. For each weather station, we have a shift and annual average temperature as well as the principal component scores. Figure 8.3 shows the correlations between the score on the relevant principal component and the two parameters estimated in the registration.

We see that the components 3 and 4 in this analysis account for very little of the original variability and have no clear interpretation. Component 1 corresponds to an increase in range between winter and summer—the effect highlighted by component 2 in the previous analysis. We see that this effect is strongly negatively correlated with annual average temperature, and mildly negatively correlated with shift. Component 2 corresponds approximately to component 4 in the previous analysis, and is the effect whereby the length of summer is lengthened relative to that of winter. This effect is positively correlated with average temperature and negatively correlated with shift.

9
Functional linear models

9.1 Introduction

So far in this book, we have concentrated on analysing the variability of a single functional variable, albeit one that may have a rather complicated structure. In classical statistics, techniques such as linear regression, the analysis of variance, and the general linear model all approach the question of how variation in an observed variable may be accounted for by other observed quantities. We now extend these general ideas to the functional context.

Linear models can be functional in terms either of the dependent variable, or of the linear mapping classically described by the design matrix, or both. In all cases, the regression coefficients or parameters are functions rather than just numbers. In this chapter, we introduce some of the ideas of functional linear modelling by extending the general linear model in one particular way, where only the parameters and the dependent variable become functional, but the design matrix remains as in the classical general linear model. We review further extensions in Section 9.5 and in subsequent chapters.

9.2 Functional linear models and functional analysis of variance

Consider the Canadian weather data introduced in Chapter 1. Monthly means for temperature and precipitation are available for each of 35 weather stations distributed across the country, and we can use the smoothing techniques of Chapter 3 to represent each 12-month record as a smooth function. Thus, two periodic functions, Temp and Prec, denoting temperature and the logarithm of precipitation, respectively, are available for each station.

In this chapter we ask how much of the pattern of annual variation of temperature is attributable to geographical area. Dividing Canada into Atlantic, Continental, Pacific and Arctic meteorological zones, we want to study the characteristic types of temperature patterns in each zone. This is basically an analysis of variance problem with four treatment groups. Multivariate analysis of variance (MANOVA) is the extension of the ideas of analysis of variance to deal with problems where the dependent variable is multivariate. Because our dependent variable is the functional observation Temp, the methodology we need is a *functional* analysis of variance, abbreviated FANOVA.

In formal terms, we have a number of stations in each group g, and the model for the kth temperature function in the gth group, indicated by $\mathtt{Temp}_{kg}$, is

$$\mathtt{Temp}_{kg}(t) = \mu(t) + \alpha_g(t) + \epsilon_{kg}(t). \tag{9.1}$$

The function μ is the grand mean function, and therefore indicates the average temperature profile across all of Canada. The terms α_g are the specific effects on temperature of being in climate zone g. To be able to identify them uniquely, we require that they satisfy the constraint

$$\sum_g \alpha_g(t) = 0 \text{ for all } t. \tag{9.2}$$

The residual function ϵ_{kg} is the unexplained variation specific to the kth weather station within climate group g.

We can define a 35×5 design matrix $\mathbf{Z}$ for this model, with one row for each individual weather station, as follows. Use the label (k, g) for the row corresponding to station k in group g; this row has a one in the first column, a one in column $g + 1$, and zeroes in the rest. Write $z_{(k,g)}$ for the 5-vector whose transpose is the (k, g) row of $\mathbf{Z}$.

We can then define a corresponding set of five regression functions β_j by setting $\beta_1 = \mu$, $\beta_2 = \alpha_1$, and so on to $\beta_5 = \alpha_4$, so that the functional vector $\beta = (\mu, \alpha_1, \alpha_2, \alpha_3, \alpha_4)'$. In these terms, the model (9.1) has the equivalent formulation

$$\mathtt{Temp}_{kg}(t) = \sum_{j=1}^{5} z_{(k,g)j}\beta_j(t) + \epsilon_{kg}(t) = \langle z_{(k,g)}, \beta(t)\rangle + \epsilon_{kg}(t) \quad (9.3)$$

or, more compactly in matrix notation,

$$\mathtt{Temp} = \mathbf{Z}\beta + \epsilon, \quad (9.4)$$

where Temp is the vector containing the 35 temperature functions, ϵ is a vector of 35 residual functions, and β is the 5-vector of parameter functions. The design matrix **Z** has exactly the same structure as for the corresponding univariate or multivariate one-way analysis of variance. The only way in which (9.4) differs from the corresponding equations in standard elementary textbooks on the general linear model is that the parameter β, and hence the predicted observations $\mathbf{Z}\beta$, are vectors of functions rather than vectors of numbers.

9.2.1 *Fitting the model*

If (9.4) were a standard general linear model, the standard least squares criterion would say that β should be chosen to minimize the residual sum of squares, which can be written $\|\mathtt{Temp} - \mathbf{Z}\beta\|^2$ in vector notation. To extend the least squares principle to the functional case, we need only reinterpret the squared norm in an appropriate way. The quantity $\mathtt{Temp} - \mathbf{Z}\beta$ is now a vector of functions. The individual components of $\mathbf{Z}\beta$ are the individual predictors $\langle z_{(k,g)}, \beta\rangle$, and so the natural least squares fitting criterion becomes

$$\mathtt{LMSSE}(\beta) = \sum_g \sum_k \int [\mathtt{Temp}_{kg}(t) - \langle z_{(k,g)}, \beta(t)\rangle]^2 \, dt. \quad (9.5)$$

Minimizing LMSSE(β) subject to the constraint $\sum_2^5 \beta_j = 0$ (equivalent to $\sum_1^4 \alpha_g = 0$) gives the least squares estimates $\hat{\beta}$ of the functional parameters μ and α_g. Section 9.4 contains some remarks about the way LMSSE is minimized in practice.

Figure 9.1 displays the resulting estimated region effects α_g for the four climatic zones, and Figure 9.2 displays the composite effects $\mu + \alpha_g$. We see that the region effects are more complex than the constant or even sinusoidal effects that one might expect:

- The Atlantic stations appear to have a temperature around 5 degrees C warmer than the Canadian average.

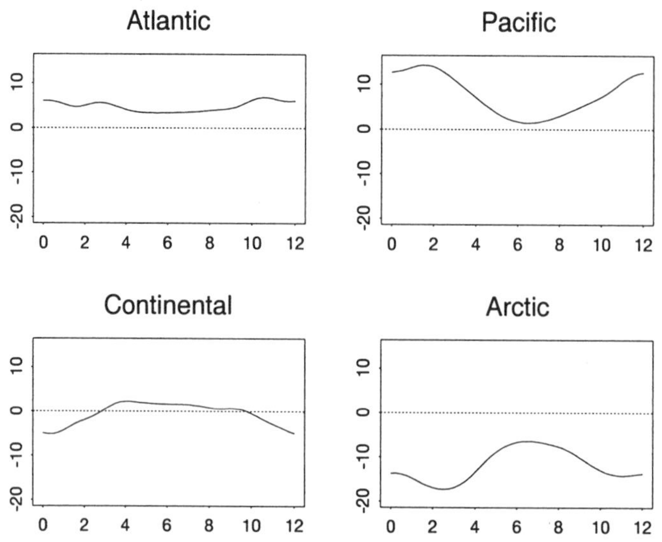

FIGURE 9.1. The region effects α_j for the temperature functions in the functional analysis of variance model $\text{Temp}_{ij}(t) = \mu(t) + \alpha_j(t) + \epsilon_{ij}(t)$. Note that the effects are required to sum to 0 for all t.

- The Pacific weather stations have a summer temperature close to the Canadian average, but are much warmer in the winter.
- The Continental stations are slightly warmer than average in the summer, but are colder in the winter by about 5 degrees C.
- The Arctic stations are certainly colder than average, but even more so in March than in January.

9.2.2 Assessing the fit

In estimating and plotting the individual regional temperature effects, we have taken our first step towards achieving the goal of characterizing the typical temperature pattern for weather stations in each climate zone. We may wish to move on and not only confirm that the total zone-specific effect α_g is nonzero, but also investigate whether this effect is substantial at a specific time t. As in ordinary analysis of variance, we look to summarize these issues in terms of error sum of squares functions LMSSE, squared correlation functions RSQ, and F-ratio functions FRATIO. It is the dependence of these quantities on t that makes the procedure different from the standard multivariate case.

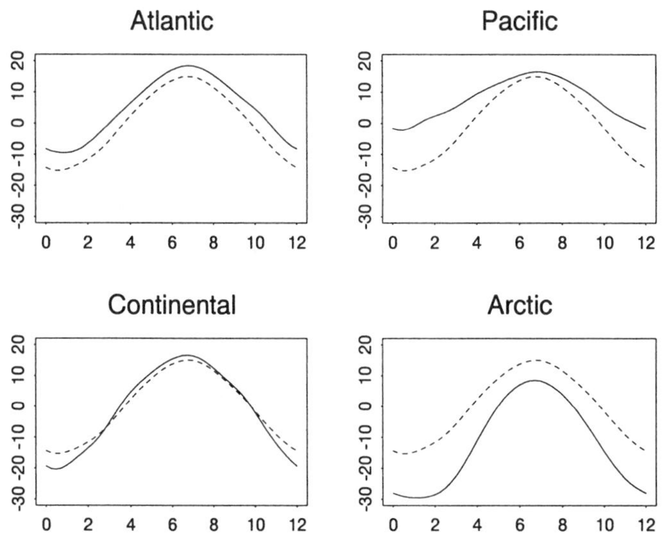

FIGURE 9.2. The estimated region temperature profiles $\mu + \alpha_j$ for the temperature functions in the functional analysis of variance model (solid curves). The dashed curve is the Canadian mean function μ.

As in the multivariate linear model, the primary source of information in investigating the importance of the zone effects α_g is the sum of squares function

$$\text{SSE}(t) = \sum_{k,g} [\text{Temp}_{kg}(t) - (\mathbf{Z}\hat{\beta})_{kg}(t)]^2. \tag{9.6}$$

This function can be compared to the error sum of squares function based on using only the Canadian average $\hat{\mu}$ as a model,

$$\text{SSY}(t) = \sum_{kg} [\text{Temp}_{kg}(t) - \hat{\mu}(t)]^2$$

and one way to make this comparison is by using the squared multiple correlation function RSQ with values

$$\text{RSQ}(t) = [\text{SSY}(t) - \text{SSE}(t)]/\text{SSY}(t). \tag{9.7}$$

Essentially, this function considers the drop in error sum of squares produced by taking climate zone into effect relative to error sum of squares without using climate zone information.

We can also compute the functional analogues of the quantities entered into the ANOVA table for a univariate analysis. For example, the mean square for error function MSE has values

$$\mathsf{MSE} = \mathsf{SSE}/\mathsf{df}(\text{error})$$

where df(error) is the degrees of freedom for error, or the sample size N less the number of mathematically independent functions β_q in the model. In this problem, the zero sum restriction on the climate zone effects α_g implies that there are four degrees of freedom lost to the model, or df(error) = 31. Similarly, the mean square for regression is the difference between SSY (or, more generally, whatever reference model we employ that is a specialization of the model being assessed) and SSE, divided by the difference between the degrees of freedom for error for the two models. In this particular application, the reference model uses one degree of freedom, so

$$\mathsf{MSR}(t) = \frac{\mathsf{SSY}(t) - \mathsf{SSE}(t)}{\mathsf{df}(\text{model})}$$

where df(model) = 3. Finally, we can compute the F-ratio function,

$$\mathsf{FRATIO} = \frac{\mathsf{MSR}}{\mathsf{MSE}}. \tag{9.8}$$

Figure 9.3 shows the two functions RSQ and FRATIO. We can see that the squared correlation is relatively high and that the F-ratio is very substantially above the 5% significance level of 2.92. It is interesting to note that the differences between the climate zones are substantially stronger in the spring and autumn, rather than in the summer and winter as we might expect.

Basically, then, most of the statistical machinery available for univariate analysis of variance is readily applicable to this functional problem. We can consider, for example, contrast functions, post-hoc multiple comparison functions, F-ratios associated with constrained estimates of region effects, and so on, essentially because the functional analysis of variance problem is really a univariate ANOVA problem for each specific value of t.

One question not addressed in the discussion of this example is an overall assessment of significance for the difference between the climate zones, rather than an assessment for each individual time t. Section 9.3.3 provides an approach to this question using simulation in the context of a different example.

An application of functional analysis of variance can be found in Ramsay, Munhall, Gracco and Ostry (1996), where variation in lip movement during the production of four syllables is analysed at the level of both position and acceleration. In the next section, we look at another application with a rich structure.

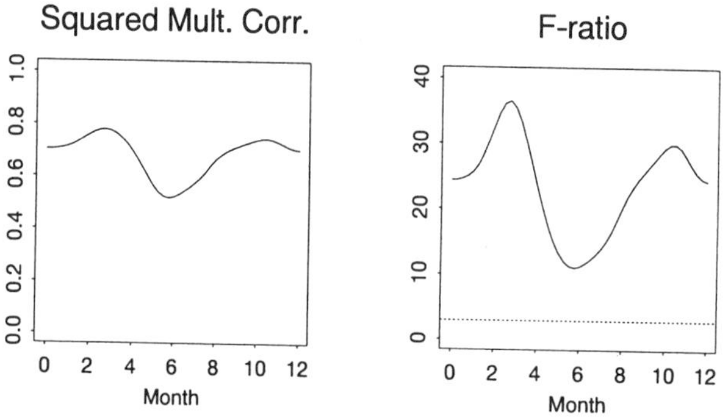

FIGURE 9.3. The left panel contains the squared multiple correlation function RSQ and the right panel the corresponding F-ratio function FRATIO. The horizontal dotted line indicates the 5% significance level for the F-distribution with 3 and 31 degrees of freedom.

9.3 Force plate data for walking horses

This section describes some interesting data on equine gait. The data were collected by Dr. Alan Wilson of the Equine Sports Medicine Centre, Bristol University, and his collaborators. Their kindness in allowing use of the data is gratefully acknowledged. The data provide an opportunity to discuss various extensions of our functional linear modelling and analysis of variance methodology. For further details of this example, see Wilson *et al.* (1996).

9.3.1 Structure of the data

The basic structure of the data is as follows. It is of interest to study the effects of various types of shoes, and various walking surfaces, on the gait of a horse. One reason for this is simply biomechanical: The horse is an animal particularly well adapted to walking and running, and the study of its gait is of intrinsic scientific interest. Secondly, it is dangerous to allow horses to race if they are lame or likely to go lame. Careful study of their gait may produce diagnostic tests of incipient lameness which do not involve any invasive investigations and may detect injuries at a very early stage, before they become serious or permanent. Thirdly, it is important to shoe horses to balance their gait, and understanding the effects of different kinds of shoe is necessary to do this. Indeed, once the normal gait of a horse is known, the measurements we describe can be used to test whether a blacksmith has shod a horse correctly, and can therefore be used as an aid in the training of farriers.

In this experiment, horses walk on to a plate about 1 metre square set into the ground and equipped with meters at each corner measuring the force in the vertical and the two horizontal directions. We consider only the vertical force. During the period that the horse's hoof is on the ground (the stance phase) the four measured vertical forces allow the instrument to measure the point of resultant vertical force. The hoof itself does not move during the stance phase, and the position of the hoof is measured by dusting the plate with sawdust or is inferred from the point of force at the end of the stride, when only the front tip of the hoof is in contact with the ground.

The vertical force increases very rapidly at the beginning of the stance phase but reduces more slowly at the end. Operationally, the stance phase is defined as starting at the moment where the total vertical force first reaches 30% of its maximum value and ending where it falls to 8% of its maximum value. For each replication, the point of force is computed for 100 time points equally spaced in this time interval.

A typical functional observation is therefore a two-dimensional function of time `Force` = (`ForceX`, `ForceY`) where t varies from 0 to 1 during the stance phase, and `ForceX`(t) and `ForceY`(t) are the coordinates of the point of force at time t. Here Y is the direction of motion of the horse, and X measures distance in a perpendicular direction towards the body of the horse. Thus the coordinates are defined as if looking at the plate from above if a left foot is being measured, but with the X direction reflected if a right foot is being measured.

The data set consists of 592 separate runs and involves 8 horses, each of which has a number of measurements on both its right and left forelimbs. The nine shoeing conditions are as follows: first, the horse is observed unshod; it is then shod and observed again; then its shoe is modified by the addition of various wedges, either building up its toe or heel or building up one side or the other of its hoof. Not every horse has every wedge applied. In the case of the toe and heel wedges, the horse is observed immediately after the wedge is fitted and one day later, after it has become accustomed to the shoe. Finally the wedges are removed and the horse is observed with a normal shoe.

Figure 9.4 shows a typical (`ForceX`, `ForceY`) plot. This realization is among the smoother curves obtained. The 100 points that are equally spaced in time are marked on the curve, and the direction of time is indicated by arrows (also evenly spaced in time). We can

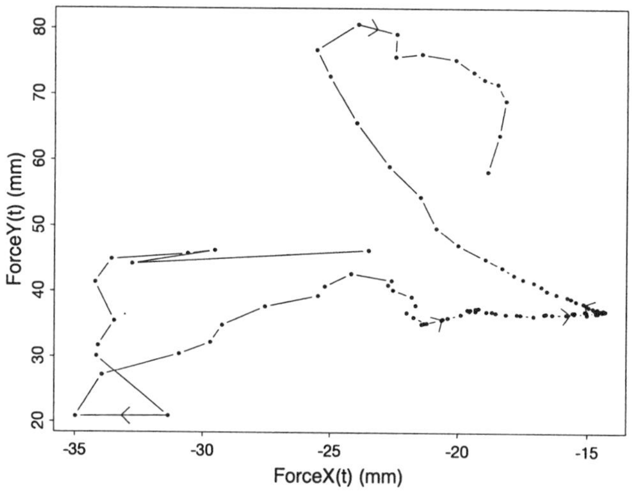

FIGURE 9.4. A typical trace of the resultant point of force during the stance phase of a horse walking onto a force plate. One hundred points equally spaced in time are indicated on the curve. The arrows indicate the direction of time.

see, not surprisingly, that the point of force moves most rapidly near the beginning and end of the stance phase. The accuracy of a point measurement was about 1 mm in each direction.

9.3.2 *A functional linear model for the horse data*

The aim of this experiment is to investigate the effects of various shoeing conditions, and particularly to study the effects of the toe and heel wedges, which change as the horse becomes accustomed to the wedge. We fit a model of the form

$$\mathtt{Force}_{ijkl} = \mu + \alpha_{ij} + \theta_k + \epsilon_{ijkl} \tag{9.9}$$

where all the terms are two-dimensional functions of t, $0 \le t \le 1$. The suffix $ijkl$ refers to the data collected for the lth observed curve for side j of horse i under condition k.

For any particular curve, use labels x and y where necessary to denote the x and y coordinates of the vector function. The following identifiability constraints are placed on the various effects, each valid for all t:

$$\sum_{i,j} \alpha_{ij}(t) = \sum_{k=1}^{9} \theta_k(t) = 0. \tag{9.10}$$

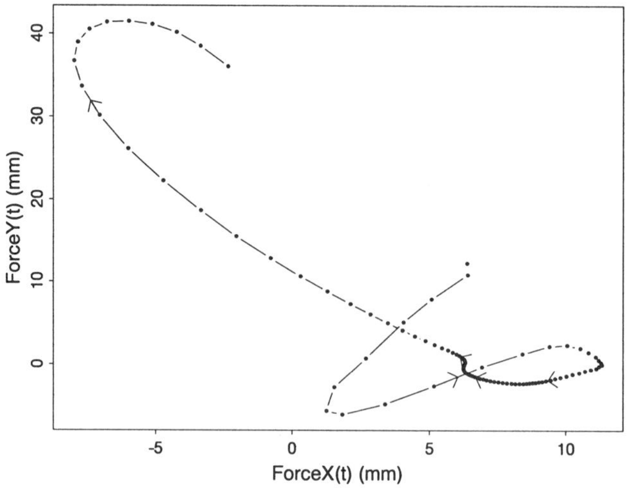

FIGURE 9.5. Estimate of the overall mean curve (μ_x, μ_y) obtained from the 592 observed point of force curves using model (9.9).

We estimate the various effects by carrying out a separate general linear model fit for each t and for each of the x and y coordinates. Since the data are observed at 100 discrete times in practice, each $\texttt{Force}_{ijkl}$ corresponds to two vectors, each of length 100, one for the x coordinates and one for the y coordinates. The design matrix relating the expected value of $\texttt{Force}_{ijkl}$ to the various effects is the same for all 200 observed values, so although the procedure involves the fitting of 200 separate models, considerable economy of effort is possible. The model (9.9) can be written as

$$\texttt{Force} = \mathbf{Z}\beta + \epsilon \tag{9.11}$$

where Force and ϵ are both vectors of length 592, each of whose elements is a two-dimensional function on $[0, 1]$. The vector β is a vector of the 26 two-dimensional functions μ, α_{ij} and θ_k, and $\mathbf{Z}$ is a 592×26 design matrix relating the observations Force to the effects β. The identifiability constraints (9.10) are incorporated by augmenting the matrix $\mathbf{Z}$ by additional rows corresponding to the constraints, and by augmenting the data vector Force by zeroes. Standard theory of the general linear model of course then gives as the estimator

$$\hat{\beta} = (\mathbf{Z}'\mathbf{Z})^{-1}\mathbf{Z}'\,\texttt{Force}. \tag{9.12}$$

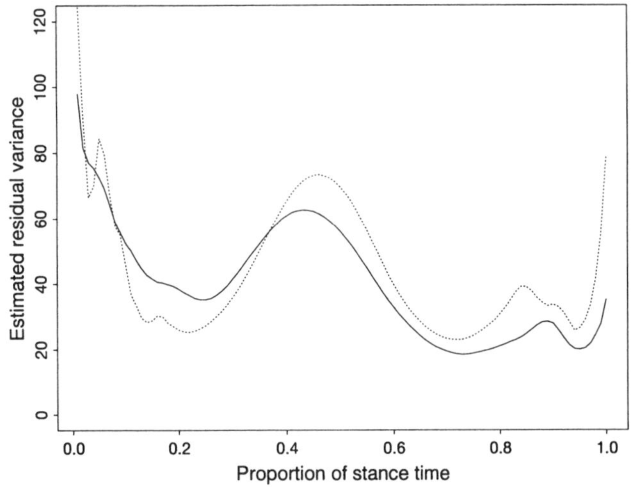

FIGURE 9.6. The estimated residual variance in the x coordinate (solid curve) and the y coordinate (dotted curve) as the stance phase progresses.

Figure 9.5 plots the estimated overall mean curve $\mu = (\mu_x, \mu_y)$ in the same way as Figure 9.4. Although the individual observations are somewhat irregular, the overall mean is smooth, even though no smoothing is incorporated into the procedure.

The general linear model fitted for each coordinate at each time point allows the calculation of a residual sum of squares, and hence an estimated residual variance, at each point. The residual variance curves $\mathtt{MSE}_x$ and $\mathtt{MSE}_y$ for the x and y coordinates are plotted in Figure 9.6. It is very interesting to note that the residual variances in the two coordinates are approximately the same size, and vary in roughly the same way, as the stance phase progresses.

9.3.3 *Effects and contrasts*

We can now explain how the linear model can be used to investigate various effects of interest. We concentrate on two specific effects, corresponding to the application of the toe wedges, and illustrate how various inferences can be drawn. In Figure 9.7, we plot the effects of the toe wedge immediately after it has been applied and the following day. The x and y coordinates of the relevant functions θ_k are plotted separately. It is interesting to note that the y effects are virtually the same in both cases: The application of the wedge has an immediate

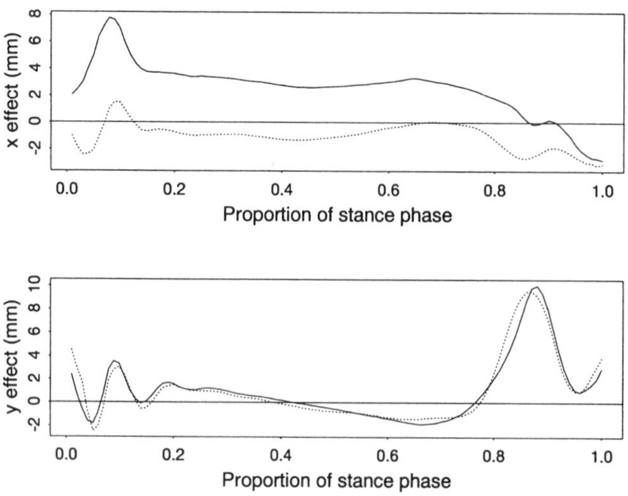

FIGURE 9.7. The effects of the application of a toe wedge, x coordinate in the lower panel and y coordinate in the lower. Solid curves are the immediate effect, and dashed curves are the effect on the following day.

effect on the way in which the point of force moves in the forward-backward direction, and this pattern does not change appreciably as the horse becomes accustomed to the wedge. The effect in the side-to-side direction is rather different. Immediately after the wedge is applied, the horse tends to put its weight to one side, but the following day the effect becomes much smaller, and the weight is again placed in the same lateral position as in the average stride.

To investigate the significance of this change, we now consider the contrast between the two effects, which shows the expected difference between the point of force function for a horse 24 hours after a toe wedge has been applied and that immediately after applying the wedge. Figure 9.8 shows the x and y coordinates of the difference of these two effects. The standard error of this contrast is easily calculated. Let u be the vector such that the estimated contrast is the vector function $\texttt{Contrast} = u'\hat{\beta}$, so that the component of u corresponding to toe wedge 24 hours after application is $+1$, that corresponding to toe wedge immediately after application is -1, and all the other components are zero. Define a by $a^2 = u'(\mathbf{Z}'\mathbf{Z})^{-1}u$. The squared pointwise standard errors of the x and y coordinates of the estimated contrasts are then $a^2\mathsf{MSE}_x$ and $a^2\mathsf{MSE}_y$, respectively. Plots of ± 2 times the relevant standard error are included in Figure 9.8. Because the

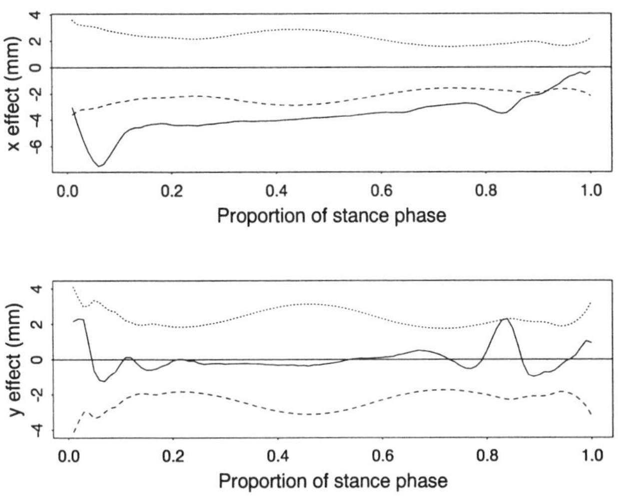

FIGURE 9.8. Solid curves are the differences between the effect of a toe wedge after 24 hours and its immediate effect. Dotted curves indicate plus and minus two estimated standard errors for the pointwise difference between the effects. The upper panel contains x coordinates and the lower the y coordinates.

degrees of freedom $(592-26+2)$ for residual variance are so large, these plots indicate that pointwise t tests at the 5% level would demonstrate that the difference in the y coordinate of the two toe wedge effects is not significant, except possibly just above time 0.8, but that the x coordinate is significantly different from zero for almost the whole stance phase.

How should we account for the correlation in the tests at different times in assessing the significance of any difference between the two conditions? We can consider the summary statistics

$$M_x = \sup_t |\mathtt{Contrast}_x(t)/a\sqrt{\mathsf{MSE}_x(t)}|$$

and

$$M_y = \sup_t |\mathtt{Contrast}_y(t)/a\sqrt{\mathsf{MSE}_y(t)}|.$$

The values of these statistics for the data are $M_x = 5.03$ and $M_y = 2.01$. A permutation-based significance value for each of these statistics was obtained by randomly permuting the observed toe wedge data for each leg of each horse between the conditions immediately after fitting of wedge and 24 hours after fitting of wedge, keeping the totals the same within each condition for each leg of each horse. The statistics M_x and M_y were calculated for each random permutation of the data. In 1000 realizations, the largest value of M_x observed was 3.57, so the observed difference in the x direction of the two conditions is highly significant. A total of 177 of the 1000 simulated M_y values exceeded the observed value of 2.01, and so the estimated p-value of this observation was 0.177, showing no evidence that the y coordinate of point of force alters its time behaviour as the horse becomes accustomed to the wedge.

9.4 Computational issues

9.4.1 The general model

In this section, we set out methods for computing the least squares estimates in the functional linear model. To be specific, we assume that Y is an N-vector of observations, with each element of some appropriate functional form. The q-vector β of parameters has elements with the same functional form as the elements of Y; thus, in the force-plate data example, the individual elements of both Y and β were themselves two-dimensional functions. We assume that $\mathbf{Z}$ is an $N \times q$ design matrix, and that the expected value of $Y(t)$ for each t is modelled as $\mathbf{Z}\beta(t)$.

Any linear constraints on the parameters β, such as the requirement in the temperature data example that the individual climate zone effects sum to zero, are expressed as $\mathbf{L}\beta = 0$ for some suitable matrix $\mathbf{L}$ with q columns. By using a technique such as the QR-decomposition, described in Section A.1.3 of the Appendix, we may also say that

$$\beta = \mathbf{C}\alpha \tag{9.13}$$

for some matrix $\mathbf{C}$.

Our aim is to minimize the least squares fitting criterion

$$\mathsf{LMSSE}(\beta) = \int \|Y(t) - \mathbf{Z}\beta(t)\|^2 \, dt \tag{9.14}$$

subject to the constraint $\mathbf{L}\beta(t) = 0$ for all t. The norm in the definition of LMSSE is an appropriate Euclidean vector norm. Assume that the constraints assure the identifiability of the parameters. In linear algebraic terms, this means that the rows of $\mathbf{Z}$ are linearly independent of the rows of $\mathbf{L}$, and that the augmented design matrix constructed by putting the rows of $\mathbf{Z}$ together with those of $\mathbf{L}$ is of full column rank q.

9.4.2 *Pointwise minimization*

Since there are no particular restrictions on the way in which $\beta(t)$ varies as a function of t, we can minimize $\texttt{LMSSE}(\beta)$ by minimizing $\|Y(t) - \mathbf{Z}\beta(t)\|^2$ individually for each t. Because the rows of $\mathbf{Z}$ and those of $\mathbf{L}$ are independent, we can compute the constrained minimum of $\texttt{LMSSE}(\beta)$ by finding the unconstrained minimum of the regularized residual sum of squares

$$\|Y(t) - \mathbf{Z}\beta(t)\|^2 + \lambda\|\mathbf{L}\beta(t)\|^2$$

for any convenient $\lambda > 0$. We can find $\hat{\beta}$ by solving the equation

$$(\mathbf{Z}'\mathbf{Z} + \lambda\mathbf{L}'\mathbf{L})\hat{\beta}(t) = \mathbf{Z}'Y(t). \tag{9.15}$$

The solution for $\hat{\beta}$ does not depend on the actual value assigned to λ, since for any positive value the incorporation of the penalty term simply imposes the constraint $\mathbf{L}\beta(t) = 0$ and does not affect the fit. The value $\lambda = 1$ usually works, but can be varied if numerical instabilities are encountered.

The most straightforward method of finding the function $\hat{\beta}$ is to find $\hat{\beta}(t)$ from (9.15) for a suitable grid of values of t, and then interpolate between these values. This was the technique used in both the temperature and force plate examples, where the grid of values was chosen to correspond with the discretization of the original data. In the case of the temperature data, only 12 values of t were used, whereas a finer grid of 100 values was appropriate for the force plate data. In either case, the fact that the same system of equations is solved repeatedly in (9.15) makes for considerable economy of numerical effort.

9.4.3 *Functional linear modelling with basis expansions*

It is often more appropriate to store the observed functions and parameters in basis expansion form, for example in terms of Fourier series or B-splines. For example, it may be that the functions are well approximated by a relatively short basis expansion. Another possibility is that the different functions may have been observed at different sets of times, and that a common basis expansion method has been used to construct the functional observations.

Suppose that our basis is a K-vector ϕ of linearly independent functions, and that the matrix $\mathbf{Y}$ gives the coefficients of the observed vector Y of functions, so that $Y = \mathbf{Y}\phi$. Thus, the (j, k) element of the matrix $\mathbf{Y}$ is the coefficient of ϕ_k in the expansion of Y_j. Expand the estimated parameter vector $\hat{\beta}$ in terms of the same basis, expressing $\hat{\beta} = \mathbf{B}\phi$ for a $q \times K$ matrix $\mathbf{B}$. We can now substitute into (9.16) to see that $\mathbf{B}$ satisfies the matrix system of linear equations

$$(\mathbf{Z}'\mathbf{Z} + \lambda\mathbf{L}'\mathbf{L})\mathbf{B} = \mathbf{Z}'\mathbf{Y}. \tag{9.16}$$

This provides a simple alternative to the discretization strategy described in Section 9.4.2.

9.4.4 *Incorporating regularization*

So far, we have merely fitted the functional linear model without imposing any regularity on the estimated parameter functions β. This normally suffices for problems where the observed functions Y are reasonably smooth. Indeed, the averaging implicit in the estimation of β from most models often leads to parameter functions that are smoother than the original observed Y.

However, there may be cases where the observed functions are so rough or rapidly varying that we wish to incorporate some regularization into the fitting of the functional linear model, over and above constraints imposed on the parameters. We comment very briefly on two possible approaches to this issue.

One simple method is to use basis expansions with a relatively small number of basis functions; this smooths the original observations by projecting them onto the space spanned by the restricted basis and hence automatically yields smoother estimates of the parameter functions β.

Another possibility is to introduce a roughness penalty into the measure of goodness of fit LMSSE. Thus, for example, we would define $\mathsf{PEN}_2(\beta)$ to be $\sum_j \mathsf{PEN}_2(\beta_j) = \sum_j \int (D^2\beta_j)^2$. Letting $\lambda \geq 0$ be a smoothing parameter, we then define

$$\mathsf{LMSSE}_\lambda(\beta) = \mathsf{LMSSE}(\beta) + \lambda \times \mathsf{PEN}_2(\beta).$$

Minimizing LMSSE subject to the constraint $\mathbf{L}\beta = 0$ then give regularized estimates of the parameters β.

Computationally, it is probably most straightforward to implement this regularization procedure by using a basis expansion. Given a basis ϕ, define $\mathbf{R}$ to be the matrix $\int (D^2\phi)(D^2\phi)'$ and let $\mathbf{J}$ be the matrix of inner products $\langle \phi, \phi' \rangle$. As in Section 9.4.3, write $\beta = \mathbf{B}\phi$ and $Y = \mathbf{Y}\phi$ for suitable matrices of coefficients $\mathbf{Y}$ and $\mathbf{B}$. Then

$$\mathsf{LMSSE}_\lambda(\beta) = \operatorname{trace}(\mathbf{Y} - \mathbf{ZB})\mathbf{J}(\mathbf{Y} - \mathbf{ZB})' + \lambda \operatorname{trace} \mathbf{BRB}'. \tag{9.17}$$

If β is unconstrained, then **B** satisfies the equation

$$(\mathbf{Z}'\mathbf{J}\mathbf{Z} + \lambda\mathbf{R})\mathbf{B} = \mathbf{Z}'\mathbf{J}\mathbf{Y}.$$

Or, if β is constrained, then from (9.13) $\mathbf{B} = \mathbf{C}\mathbf{A}$ for some unknown matrix **A**, and substituting this relationship in (9.17) yields

$$(\mathbf{C}'\mathbf{Z}'\mathbf{J}\mathbf{Z}\mathbf{C} + \lambda\mathbf{C}'\mathbf{R}\mathbf{C})\mathbf{A} = \mathbf{C}'\mathbf{Z}'\mathbf{J}\mathbf{Y}.$$

9.5 General structure

We close this chapter by briefly discussing some other possible functional linear modelling problems. The equation (9.4) developed in this chapter looks exactly the same as the standard multivariate general linear model. A vector of parameters β is mapped by a matrix **Z** into a vector of predictors of the observations. In the examples we have discussed, the key difference is that the parameters and the individual observations are functions rather than just numbers.

We can step back and ask what is the general structure of these problems. The matrix **Z** defines a *linear transformation* from the parameter space $\mathcal{B}$ to the observation space $\mathcal{Y}$, such that for any β in $\mathcal{B}$, the prediction $\hat{Y}$ of the observation Y is equal to $\mathbf{Z}\beta$. In the classical general linear model, the members of the parameter space $\mathcal{B}$ are vectors β. In the functional linear model (9.3) they are vector-valued functions β, as are the members Y of $\mathcal{Y}$, but the linear transformations considered are all of the form $\mathbf{Z}\beta$ for matrices **Z**.

In some circumstances, it is appropriate to consider more general linear transformations Z of the parameter space into the observation space. In every case, we regard Z as the *model* that gives the predicted observations in terms of the parameters. In Chapter 10, we consider problems where the observations y_i are scalars rather than functions, but the parameters β_j are functional. Our example is the prediction of *total* annual log precipitation from the annual pattern of temperature. In this case, the model Z is a linear mapping from a space of vector-valued functions β to an ordinary space of multivariate observations y.

In Chapter 11 we return to problems where both the response y and the parameter β are functional. However, rather than consider the relatively simple situation where $\mathcal{B}$ is mapped to $\mathcal{Y}$ by a matrix **Z** of numbers, we look at problems where Z is a much more general functional operator.

10

Functional linear models for scalar responses

10.1 Introduction: Functions as predictors

In this chapter, we consider a linear model defined by a set of functions, but where the response variable is scalar or multivariate. This contrasts with Chapter 9, where the responses and the parameters were functional, but, because of the finite and discrete covariate information, the linear transformation from the parameter space to the observation space was still specified by a design *matrix* $\mathbf{Z}$ as in the conventional multivariate general linear model.

We begin by recalling some aspects of ordinary linear regression. Suppose $y_1, \ldots, y_N$ are observations of a response variable at values $x_1, \ldots, x_N$ of a multivariate covariate x of dimension p. Linear regression, of course, then fits a model of the form

$$y_i = \alpha + \sum_j \beta_j x_{ij} + \epsilon_i = \alpha + \langle \beta, x_i \rangle + \epsilon_i \tag{10.1}$$

where ϵ_i is a residual or disturbance term. The usual fitting method is by least squares, estimating α and β to minimize the residual sum of squares

$$\mathsf{LMSSE}(\alpha, \beta) = \sum_{i=1}^{N} (y_i - \alpha - \langle \beta, x_i \rangle)^2. \tag{10.2}$$

We can now consider a functional extension of linear regression, where the prediction of the scalar values y_i is based on functions

x_i. This problem is of interest in its own right, and also raises issues that become relevant when we consider more complicated problems in subsequent chapters. For illustration, let us investigate to what extent total annual precipitation for Canadian weather stations can be predicted from the pattern of temperature variation through the year. To this end, let y_i be the logarithm of total annual precipitation at weather station i, and let $x_i = \texttt{Temp}_i$ be its temperature function. The interval $\mathcal{T}$ containing index t is either [0,12] or [0,365], depending on whether monthly or daily data, respectively, are involved.

To extend the idea of ordinary linear regression, we must replace the parameter β in equation (10.1) by a function β, so that the model takes the form

$$y_i = \alpha + \int_0^T x_i(s)\beta(s)\,ds + \epsilon_i = \alpha + \langle x_i, \beta \rangle + \epsilon_i. \tag{10.3}$$

In classical linear regression it is usually assumed that the covariates x are observed without error; another way of expressing this assumption is to say that the analysis is carried out conditional on the covariates having the values actually observed. We shall make a similar assumption when considering regression models with functional covariates, and consider models based on the observed covariate functions. It may be that these have been preprocessed to obtain smooth covariates. For functional analytic reasons beyond the scope of our present treatment, such smoothing will not alleviate the need for regularization within the functional regression demonstrated in the next section.

10.2 A naïve approach: Functional interpolation

We now attempt to proceed along the same lines as standard linear regression, but we shall see that some modifications are needed to give meaningful results. Suppose, naïvely, that we attempt to estimate the parameters in (10.3) by minimizing the residual sum of squares, noting that with the use of inner product notation, this is expressed as

$$\begin{aligned} \texttt{LMSSE}(\alpha, \beta) &= \sum_{i=1}^{N} (y_i - \alpha - \langle x_i, \beta \rangle)^2 \\ &= \|y - \alpha - \langle x, \beta \rangle\|^2 \end{aligned} \tag{10.4}$$

where x is the N-vector of covariate functions $(x_1, \ldots, x_N)'$.

For the temperature and precipitation data, it turns out that we can find a value α and a function β which yield a residual sum of squares

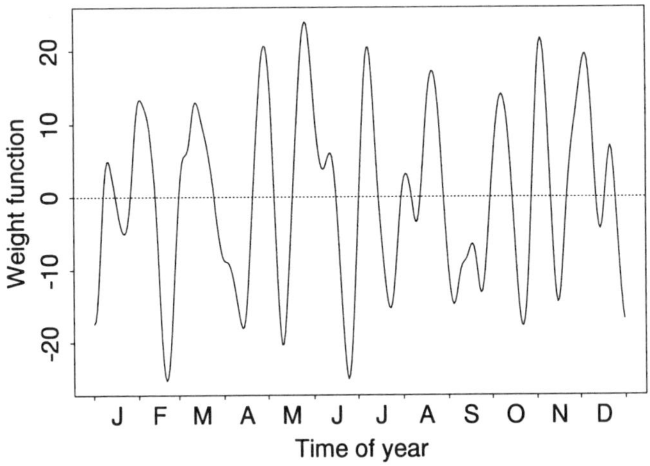

FIGURE 10.1. The weight function β that allows perfect prediction of log total annual precipitation from observed annual pattern of temperature.

of zero, and hence a perfect fit in the model (10.3) with no error at all! Not only is it somewhat counterintuitive that temperature patterns can predict overall precipitation perfectly, but the estimated regression coefficient function $\hat{\beta}$ shown in Figure 10.1 does not seem to convey any useful information.

A moment's reflection yields the reason for this perfect fit. Perhaps we can most easily explain this by reference to the high-dimensional multivariate problem presented by the daily data. Suppose x_{ij} is the entry for the temperature at station i on day j, and we wish to predict y_i by

$$\hat{y}_i = \alpha + \sum_{j=1}^{365} x_{ij}\beta_j, \; i = 1, 2, \ldots, 35.$$

We can view this as a finely discretized version of the functional model being considered. This is a system of 35 equations with 366 unknowns. Even if the coefficient matrix is of full rank, there are still infinitely many sets of solutions, all giving a perfect prediction of the observed data.

Returning to the functional model (10.3), we now see that the regression coefficient function β is underdetermined on the basis of any finite sample (x_i, y_i), because, essentially, we have an infinite number of parameters $\beta(s)$ and only a finite number of conditions

$y_i = \alpha + \langle x_i, \beta \rangle$ to satisfy. Usually it is possible to find $\hat{\alpha}$ and $\hat{\beta}$ to reduce the residual sum of squares (10.4) to zero. Furthermore, if β^* is any function satisfying $\langle x_i, \beta^* \rangle = 0$ for $i = 1, \ldots, N$, then adding β^* to $\hat{\beta}$ does not affect the value of the residual sum of squares.

Since the space of functions satisfying (10.3) is infinite-dimensional no matter how large our sample size N is, minimizing the residual sum of squares cannot, of itself, produce a meaningful or consistent estimator of the parameters β in the model (10.3). Consequently, to provide a good estimator or even just identify $\hat{\beta}$ uniquely, we must use some method of *regularization*, and this is discussed in the following sections.

10.3 Regularization by discretizing the function

In the weather data example, a possible approach is to reduce the number of unknowns in problem (10.2) by considering the temperatures on a coarser time scale. It is unlikely that overall precipitation is influenced by details of the temperature pattern from day to day, and so, for example, we could investigate how the 12-vectors of monthly average temperatures can be used to predict total annual precipitation. If $\mathbf{X}$ is the 35×12 matrix containing these values, we can then fit a model of the form $\hat{y} = \alpha + \mathbf{X}\beta$, where $\hat{y}$ is the vector of values of log annual precipitation predicted by the model, and β is a 12-vector of regression parameters. Since the number of parameters to be estimated is now only 13, and thus less than the number of observations $N = 35$, we can use a standard multiple regression to fit the model by least squares.

We can summarize the fit in terms of the conventional $R^2 = 1 - \mathsf{SSE/SSY}$ measure, and this is 0.84, indicating a rather successful fit, even taking into account 12 degrees of freedom in the model. The corresponding F-ratio is 9.8 with 12 and 22 degrees of freedom, and is significant at the 1% level. The standard error estimate is 0.34, as opposed to the standard deviation of the dependent variable of 0.69.

Figure 10.2 presents the estimated regression function β, obtained by interpolating the individual estimated coefficients $\hat{\beta}_j$ as marked on the figure. It is not easy to interpret this function directly, although it clearly places considerable emphasis on temperature in the months of March, April, September and December. The lack of any very clear interpretation indicates that this problem raises statistical questions beyond the formal difficulty of fitting an underdetermined model. Certainly, the model uses up a rather large proportion of the 35 degrees

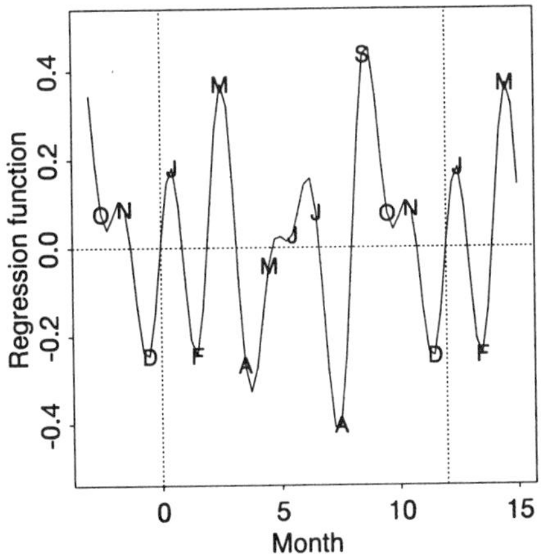

FIGURE 10.2. The regression function β for the approximation of centred annual mean log precipitation by the temperature profiles for the Canadian weather stations.

of freedom available in the data. Might it not be wise to impose even more smoothness on β?

10.4 Regularization using basis functions

To reduce the degrees of freedom in the model still further, we could average over longer time periods than one month. This would introduce a considerable degree of arbitrariness in terms of the starting point of the year. A preferable approach for the periodic data we are considering is to expand the temperature curves in a Fourier basis. Of course, we can employ a similar approach for nonperiodic data using some other suitable basis, and this is discussed in Section 10.4.3.

10.4.1 Re-expressing the model and data: Fourier bases

Let $\phi_0, \phi_1, \ldots$ be the Fourier basis given in (3.10) for $\mathcal{T} = [0, T]$, and let M be the largest number of these basis functions that we might propose to use. For the monthly and daily temperature data, for example, M would be 12 and 365, respectively. More generally, we may choose to truncate the expansion at some suitably large K that does not entail any significant loss of information.

Now expand

$$x_i = \sum_\nu c_{i\nu}\phi_\nu = c_i'\phi \tag{10.5}$$

where $c_{i\nu}$ is the Fourier coefficient for observation i and basis function ϕ_ν, and at the same time let b be the vector of Fourier coefficients of the regression function β, implying that

$$\beta = \sum_\nu b_\nu\phi_\nu = b'\phi. \tag{10.6}$$

The Parseval identity for Fourier series states that

$$\langle x_i, \beta\rangle = \sum_\nu c_{i\nu}b_\nu.$$

We can further simplify notation by defining the $(M+1)$-vector $\zeta = (\alpha, b_0, \dots, b_{M-1})'$ and defining the coefficient matrix $\mathbf{Z}$ to be the $N \times (M+1)$ matrix $\mathbf{Z} = [\mathbf{1} \;\; \mathbf{C}]$ where $\mathbf{C}$ contains the coefficients $c_{i\nu}$. Then the model (10.1) becomes simply

$$\hat{y} = \mathbf{Z}\zeta \tag{10.7}$$

and the least squares estimate of the augmented parameter vector ζ is the solution of the equation

$$\mathbf{Z}'\mathbf{Z}\hat{\zeta} = \mathbf{Z}'y. \tag{10.8}$$

10.4.2 Smoothing by basis truncation

One very natural method of regularization is to truncate the basis by choosing a value $K < M$, and setting all but the first $K+1$ coefficients ζ_ν to be zero in both (10.5) and (10.6). We can then fit ζ by least squares, and the problem is now a standard multiple regression problem. Once the estimate of the weight function ζ is available, we can retrieve β itself by substituting into (10.6), in practice by using an inverse fast Fourier transform.

Figure 10.3 shows the result of carrying out this procedure for the daily weather data with varying numbers K of basis functions. The choice $K = 12$ is intended to correspond to the same amount of discretization as using monthly average data, and we can see that the weight function is similarly uninformative. To obtain results more likely to be meaningful, we have to use a much smaller number of basis functions, and, by considering the graphs for $K = 4$ and $K = 3$,

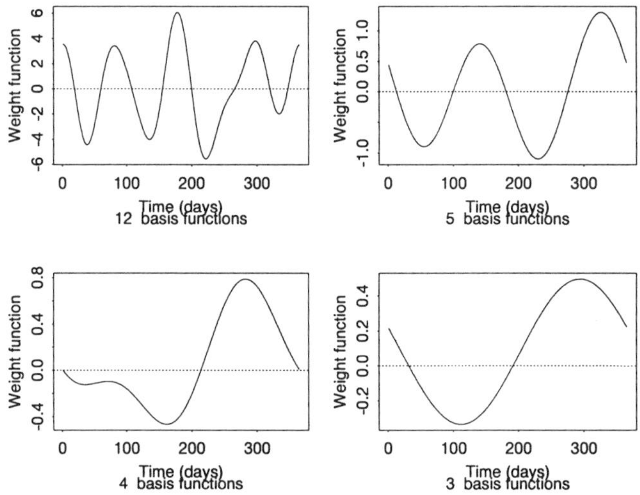

FIGURE 10.3. Estimated regression weight functions β using $K = 12, 5, 4$ and 3 basis functions.

it appears that a predictor for high precipitation is relatively high temperature towards the end of the year.

But the model complexity increases in discrete jumps as K varies from three to five, and we might want finer control. Also, to obtain reasonable results, β must be rigidly constrained to lie in a low-dimensional parametric family, and we may worry that we are missing important features in β as a consequence. Section 10.5.1 develops a more flexible approach making use of a roughness penalty method.

10.4.3 *Re-expressions with other bases*

There is no need for us to restrict attention to Fourier bases. For most bases it is not necessarily appropriate to reduce the dimensionality by truncating the basis expansion; instead, the required dimensionality is specified in advance. For example B-splines on a reasonably fine mesh provide a good basis for expanding nonperiodic functions. But note that using a coarse mesh to reduce the dimensionality does not correspond to merely truncating the fine mesh expansion.

For any given K, suppose that $\phi = (\phi_1, \ldots, \phi_K)'$ is a basis and that regression function β has the expansion

$$\beta = \sum_{\nu=1}^{K} b_\nu \phi_\nu = b' \phi.$$

Correspondingly, expand the covariates as

$$x_i = \sum_{\nu=1}^{K} c_{i\nu}\phi_\nu = c_i'\phi.$$

To allow for bases that are not necessarily orthonormal (such as B-splines), define $\mathbf{J}$ to be the matrix $\mathbf{J} = \int \phi\phi'$ with entries

$$\mathbf{J}_{jk} = \int \phi_j(s)\phi_k(s)\,ds = \langle \phi_j, \phi_k \rangle. \tag{10.9}$$

Now

$$\langle x_i, \beta \rangle = \sum_{j=1}^{K}\sum_{\nu=1}^{K} c_{ij}\mathbf{J}_{j\nu}b_\nu,$$

and the coefficient matrix $\mathbf{Z}$ is now defined by $\mathbf{Z} = [1 \quad \mathbf{CJ}]$. With this modification, we can now proceed in the same way as for the Fourier basis. This results as before in (10.7) and the estimate given by (10.8).

10.5 Regularization with roughness penalties

10.5.1 Penalizing the residual sum of squares

The estimated function $\hat{\beta}$ in Figure 10.1 illustrates that fidelity to the observed data, as measured by the residual sum of squares, is not the only aim of the estimation. The roughness penalty approach makes explicit the complementary, possibly even conflicting, aim of avoiding excessive local fluctuation in the estimated function. To this end, given any periodic twice differentiable function β, we can define the penalized residual sum of squares

$$\begin{aligned} \text{PENSSE}_\lambda(\alpha, \beta) &= \sum_{i=1}^{N}[y_i - \alpha - \langle x_i, \beta \rangle]^2 + \lambda \int_{\mathcal{T}} [D^2\beta(s)]^2\,ds \\ &= \|y - \alpha - \langle x_i, \beta \rangle\|^2 + \lambda \|D^2\beta\|^2. \end{aligned} \tag{10.10}$$

As in Chapter 4, the integrated squared second derivative or curvature $\int (D^2\beta)^2$ quantifies the rapid variation in β, and the smoothing parameter $\lambda > 0$ controls the trade-off between roughness and infidelity to the observed data. Sections 10.6 and 10.7 discuss the algorithmic aspects of minimizing (10.10).

Figure 10.4 illustrates the effect of varying the smoothing parameter. For small values of λ, the estimate is too variable for us to draw any meaningful conclusions. The feature of the model that is most sensitive

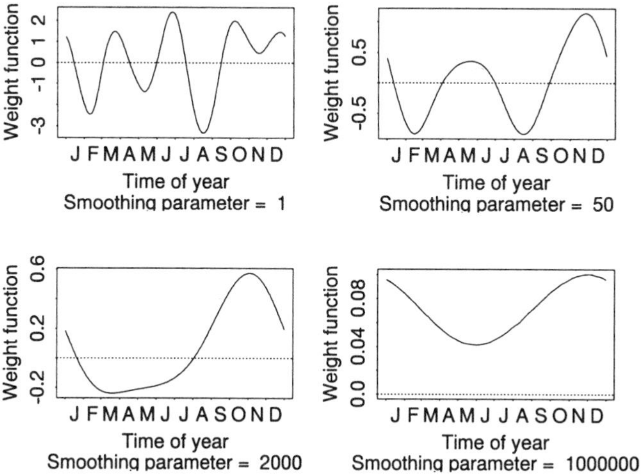

FIGURE 10.4. Weight functions estimated for various values of the smoothing parameter.

to the choice of λ is the overall effect of the first eight months of the year, in contrast to the effect mainly attributable to the months around February and August. Ultimately, it is impossible to give a conclusive answer to this question merely on the basis of a small data set. Section 10.7 discusses the behaviour of the estimate for very large λ.

We can choose the smoothing parameter λ either subjectively or by an automatic method such as cross-validation. To apply the cross-validation paradigm in this context, let $\alpha_\lambda^{(-j)}$ and $\beta_\lambda^{(-j)}$ be the estimates of α and β obtained by minimizing the penalized residual sum of squares based on all the data except (x_i, y_i). We can define the cross-validation score as

$$\text{CV}(\lambda) = \sum_{j=1}^{N} (y_j - \alpha_\lambda^{(-j)} - \langle x_j, \beta_\lambda^{(-j)} \rangle)^2, \tag{10.11}$$

and minimizing $\text{CV}(\lambda)$ over λ gives an automatic choice of λ. In practice, there are efficient algorithms for calculating the cross-validation score, and Section 10.8 discusses these.

Choosing λ by cross-validation gives the curve $\hat{\beta}$ shown in Figure 10.5, and the λ value is 650. The plot of the cross-validation score $\text{CV}(\lambda)$ given in Figure 10.6 would suggest that we might also tolerate values of λ as high as about 50,000.

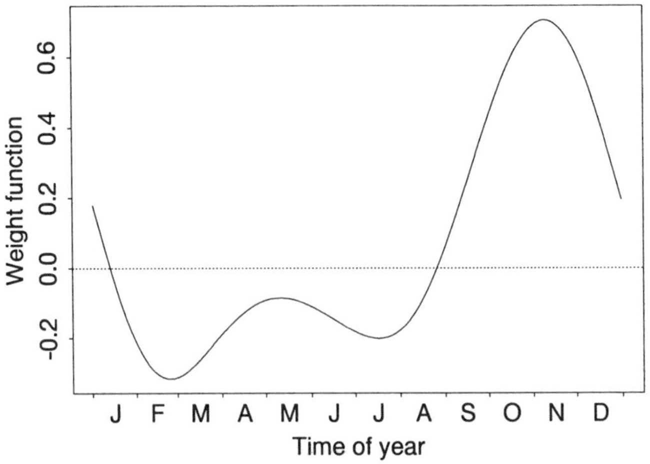

FIGURE 10.5. The estimated weight function for predicting the log total annual precipitation from the daily temperature pattern. The estimate was constructed by the roughness penalty method, with the smoothing parameter $\lambda = 650$ chosen by cross-validation.

Higher precipitation is associated with higher temperatures in the last three months of the year and with lower temperatures in spring and early summer. In effect, the estimated regression function $\hat{\beta}$ defines a *contrast* between average temperatures in October to November and average temperatures in the longer period February to July, so that rainy stations have a relatively warm autumn and cool spring, relative to the spring-autumn differential for the average station. In fact, this situation is typical of weather stations in coastal regions, where the influence of the sea is to retard the seasons, as we have already seen in Chapter 5.

In Figure 10.7, we have plotted the observed values y_i against the fitted values $\hat{y}_i$ obtained using this functional regression. This simple regression diagnostic seems to confirm the model assumptions. Section 10.8 describes another diagnostic plot.

10.5.2 Connections with nonparametric regression

Some connections with nonparametric regression, as discussed for example by Green and Silverman (1994), may be instructive. In the ordinary regression case, the assumption of linearity is sufficient to

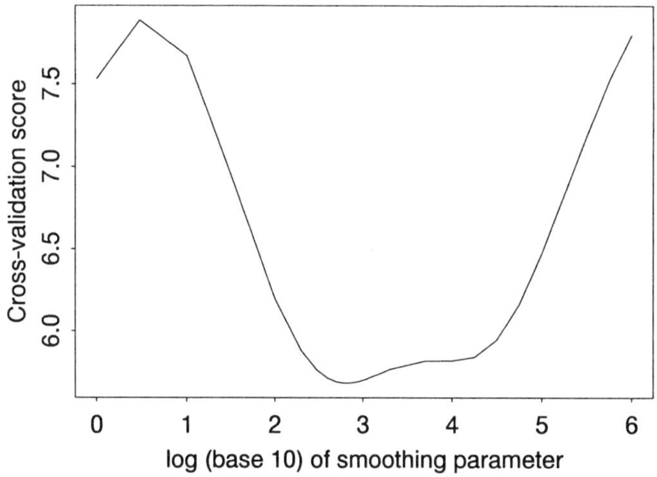

FIGURE 10.6. The cross-validation score function $\mathrm{CV}(\lambda)$ for fitting log annual precipitation by daily temperature variation. The logarithm of the smoothing parameter is taken to base 10. The minimum is at $\lambda = 650$, and values lower than 500 or larger than 50,000 seem unreasonable.

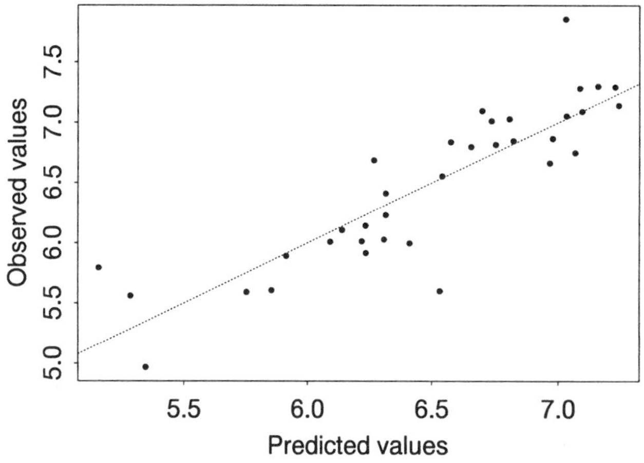

FIGURE 10.7. Observed values y_i of log annual precipitation plotted against the values $\hat{y}_i$ predicted by the functional regression model with the smoothing parameter chosen by cross-validation. The straight line on the figure is the graph observed value = predicted value.

ensure that a model of the form (10.1) can be fitted by least squares. However, if we relax this assumption and consider the more general model

$$y = g(x) + \epsilon$$

for some function g, then a development parallel to that of Sections 10.2 to 10.5.1 can be set out, with the following features:

- We can find, in general, an infinite number of possible functions g that interpolate the observed data.
- All these interpolating curves yield zero residual sum of squares $\sum_i [y_i - g(x_i)]^2$.
- In suitable circumstances, minimizing a penalized sum of squares

$$\sum_i [y_i - g(x_i)]^2 + \lambda \|D^2 g\|^2$$

  can give a good estimate $\tilde{g}$. We can choose the smoothing parameter λ by a suitable cross-validation approach.
- As $\lambda \to 0$, $\tilde{g}$ tends towards the smoothest interpolant to the data, but when $\lambda \to \infty$, the limiting case is the ordinary linear regression fit, $g(x_i) = \alpha + \beta x_i$.

A notable difference between functional regression and ordinary regression is that in the functional regression case, the unconstrained linear estimator interpolates the given data in the sense of predicting them precisely, whereas in ordinary regression, it can only approximate them as a rule. Thus, in the functional context, linearity no longer requires a sacrifice in fitting power.

10.6 Computational aspects of regularization

In this section, we consider various approaches to calculating functional regression weight functions and cross-validation scores. The choice of which method is best often depends on the size of the problem. Modern computing power means that for problems of moderate size (in whatever respect), it is no longer necessary to be unduly concerned with very fast computation, but some applications involve large data sets and/or functions observed at very high sampling rates. In both of these cases we must take care to avoid unnecessarily burdensome computations.

10.6.1 Direct discretization and calculation

We can approximately minimize the penalized residual sum of squares (10.10) by discretizing the covariate curves x_i and the parameter function β onto a fine grid. At the same time we can approximate the roughness penalty $\lambda\|D^2\beta\|^2$ by a multiple of the sum of squares of second differences, and the integrals $\int x_i\beta$ by sums. The minimization is then of a high-dimensional quadratic form, and can be carried out by standard numerical methods.

10.6.2 Full basis expansion

A basis function approach has appeal because it is especially simple to apply, and moreover some problems in any case suggest a particular choice of basis. The periodic nature of the temperature and precipitation data, for example, seems naturally to call for the use of a Fourier series basis. Our first strategy is therefore to represent the regularized fitting problem in terms of a basis function expansion, and then to apply the concept of regularization to this representation.

Suppose that we expand the covariate functions x_i and the regression function β to M terms relative to a basis functions ϕ_ν, as in (10.5) and (10.6) above. Define a matrix $\mathbf{K}$ to have entries

$$\mathbf{K}_{jk} = \int D^2\phi_j(s)D^2\phi_k(s)\,ds = \langle D^2\phi_j, D^2\phi_k\rangle. \tag{10.12}$$

In the Fourier case, note that $\mathbf{K}$ is diagonal, with diagonal elements ω_k^4 as in Section 7.4.1. In general, the penalized residual sum of squares can be written

$$\begin{aligned}\text{PENSSE}_\lambda(\alpha,\beta) &= \sum_i (y_i - \alpha - \sum_\nu c_{i\nu}b_\nu\phi_\nu)^2 + \lambda\int\{\sum_\nu b_\nu D^2\phi_\nu(s)\}^2 ds \\ &= \|y - \alpha - \mathbf{C}b\|^2 + \lambda b'\mathbf{K}b. \end{aligned}\tag{10.13}$$

As before, we deal with the additional parameter α by defining the augmented vector $\zeta = (\alpha, b)'$, and at the same time use $\mathbf{Z}$ as the $N\times(M+1)$ coefficient matrix $[1\ \mathbf{C}]$ if we are using a Fourier basis; more generally $\mathbf{Z} = [1\ \ \mathbf{CJ}]$, with $\mathbf{J}$ defined as in (10.9). Finally, let the penalty matrix $\mathbf{K}$ be augmented by attaching a leading column and row of $M+1$ zeros to yield $\mathbf{K}_0$. In terms of these augmented arrays, the expression (10.13) further simplifies to

$$\text{PENSSE}_\lambda(\zeta) = \|y - \mathbf{Z}\zeta\|^2 + \lambda\zeta'\mathbf{K}_0\zeta. \tag{10.14}$$

It follows that the minimizing value $\hat{\zeta}$ satisfies

$$(\mathbf{Z}'\mathbf{Z} + \lambda\mathbf{K}_0)\hat{\zeta} = \mathbf{Z}'y. \tag{10.15}$$

If the number M of basis functions is not too large, then we can solve this equation directly to find $\hat{\alpha}$ and $\hat{\beta}$, but this may be prohibitive if M is as large as 365, as it would be for the daily temperature data.

10.6.3 Restricting the basis expansion

Typically, regularizing in this way produces a relatively smooth function β, and in the Fourier case the diagonal entries of $\mathbf{K}$ increase rapidly. A useful approach is to work within a lower-dimensional problem by using a moderate number K of basis functions. In the Fourier case, this corresponds to forcing the coefficients of $\hat{\zeta}$ beyond some point $K + 1$ to be zero. This reduces the size of the system of equations (10.15).

However, in contrast to Section 10.4.2, the idea is not to use K as a smoothing parameter. Instead, the regularization is still primarily controlled by the parameter λ, and the dimension reduction is purely a numerical device to reduce the computation without substantially altering the actual result. In practice, values of K around 20 or 30 give good results, except possibly for very small values of the smoothing parameter and very large data sets.

10.7 The direct penalty method for computing β

We now turn to a more direct way of using the roughness penalty approach that computes $\hat{\beta}$ direction without using basis functions. Our first task is to show how we can set up this approach as a two-stage process, firstly minimizing a simple quadratic expression to obtain the vector of values $\hat{y}$ approximating the data vector y, and secondly computing the smoothest linear functional interpolant of these values.

10.7.1 Functional interpolation

We have already seen that the observed data can, in general, be fitted exactly by an infinite number of possible parameter choices (α, β). In some contexts, it may be of interest to define a functional interpolant $(\tilde{\alpha}, \tilde{\beta})$ to the given data by the smoothest parameter function choice that fits the data exactly. In any case, we need to consider this problem in defining the technique used to compute the estimate for β in Figure 10.5. Therefore, we require that estimate $(\tilde{\alpha}, \tilde{\beta})$ minimizes $\|D^2\beta\|^2$, subject to the N constraints

$$y_i = \tilde{\alpha} + \langle x_i, \tilde{\beta} \rangle. \tag{10.16}$$

The functional interpolant is the limiting case of the regularized estimator of Section 10.5.1 as $\lambda \to 0$. In fact, the curve $\tilde{\beta}$ resulting from interpolating the weather data is the curve shown in Figure 10.1.

We can consider this minimization problem (10.16) as a way of quantifying the roughness or irregularity of the response vector y relative to the observed functional covariates x_i. More generally, if $z_1, \dots, z_N$ is any sequence of values, then we can define the roughness of z relative to the functional covariates x_i as the roughness of the smoothest function β_z such that

$$z_i = \alpha_z + \langle x_i, \beta_z \rangle$$

for all i, for some constant α_z. This method of defining the roughness of a variate z_i will be of considerable conceptual and practical use later.

10.7.2 The two-stage minimization process

Section 10.7.3 shows that we can define an order N matrix $\mathbf{R}$ such that the roughness of a variate z can be expressed as the quadratic form

$$\int [D^2\beta(s)]^2 \, ds = z'\mathbf{R}z.$$

Assuming this to be true for the moment, we can conceptualize the solution of the smoothing problem by dividing the minimization of the penalized residual sum of squares into two stages:

Stage 1: Find predicted values $\hat{y}$ that minimize $\text{PENSSE}_\lambda(\hat{y}) = \sum_i (y_i - \hat{y}_i)^2 + \lambda \hat{y}'\mathbf{R}\hat{y}$, the solution to which is

$$\hat{y} = (\mathbf{I} + \lambda\mathbf{R})^{-1} y.$$

Stage 2: Find the smoothest linear functional interpolant (α, β) satisfying

$$\hat{y}_i = \alpha + \int x_i(s)\beta(s) \, ds. \tag{10.17}$$

The following argument shows this two stage-procedure indeed minimizes $\text{PENSSE}_\lambda(\alpha, \beta)$. Write the minimization problem as one of first minimizing $\text{PENSSE}_\lambda(\alpha, \beta)$ as a function of (α, β) but with $\hat{y}$ fixed, and then minimizing the result with respect to $\hat{y}$. Formally, this is expressed as

$$\begin{aligned} &\min_{\hat{y}} \; [\min_{\alpha,\beta}\{\text{PENSSE}_\lambda(\alpha, \beta)\}\;] \\ &= \min_{\hat{y}}\{\sum (y_i - \hat{y}_i)^2 + \lambda \min_{\beta} \int [D^2\beta(s)]^2 \, ds\}, \qquad (10.18) \end{aligned}$$

where the inner minimizations over α and β are carried out keeping fixed the values of the linear functionals $\hat{y}_i$, as defined in (10.17).

But according to our assumption, these inner minimizations yield (α, β) as the smoothest functional interpolant to the variate $\hat{y}$, so we may now write the equation as

$$\text{PENSSE}_\lambda(\alpha, \beta) = \min_{\hat{y}}\{\sum(y_i - \hat{y}_i)^2 + \lambda\hat{y}'\mathbf{R}\hat{y}\}. \tag{10.19}$$

For a moment setting aside the question of how $\mathbf{R}$ is defined, one of the advantages of the roughness penalty approach to regularization is that it allows this conceptual division to be made, in a sense uncoupling the two aspects of the smoothing procedure. However, it should not be forgotten that the roughness penalty is used in constructing the matrix $\mathbf{R}$, and so the functional nature of the covariates x_i, and the use of $\int(D^2\beta)^2$ to measure the variability of the regression coefficient function β, are implicit in both stages set out above.

We can think of the two-stage procedure in two ways: first as a practical algorithm in its own right, and second as an aid to understanding and intuition. In subsequent chapters we shall see that it has wider implications than those discussed here.

To use the algorithm in practice, it is necessary to derive the matrix $\mathbf{R}$, and we now show how to do this.

10.7.3 Functional interpolation revisited

In this section, we present an algorithmic solution to the linear functional interpolation problem presented in Stage 2 of the two-stage procedure set out in Section 10.7.2. The aim is to find the smoothest functional interpolant $(\tilde{\alpha}, \tilde{\beta})$ to a specified N-vector $\hat{y}$ relative to the given covariates $x_i, i = 1, \ldots, N$. For practical purposes, our algorithm is suitable for the case where the sample size N is moderate, and matrix manipulations of $N \times N$ matrices do not present an unacceptable computational burden.

Let the matrix $\mathbf{Z}$ be defined in terms of the functional covariates x_i as described in Section 10.4. In terms of basis expansions, we wish to solve the problem

$$\min\{\zeta'\mathbf{R}\zeta\} \text{ subject to } \mathbf{Z}\zeta = \hat{y}. \tag{10.20}$$

First we define some more notation. By rotating the basis if necessary, assume that the first M_0 basis functions ϕ_ν span the space of all functions f with roughness $\int(D^2f)^2 = 0$. In the Fourier case, this is true without any rotation: The only periodic functions with zero roughness are constants, so $M_0 = 1$, and the basis ϕ_ν consists of just the constant function.

Let $\mathbf{K}_2$ be the matrix obtained by removing the first M_0 rows and columns of $\mathbf{K}$. Then $\mathbf{K}_2$ is strictly positive-definite, and the rows and columns removed are all zeroes. In the Fourier case, $\mathbf{K}_2$ is diagonal.

Corresponding to the above partitioning, let $\mathbf{Z}_1$ be the matrix of the first M_0+1 columns of $\mathbf{Z}$, and let $\mathbf{Z}_2$ be the remaining columns. Defining $\mathbf{P}$ to be the $N \times N$ projection matrix $\mathbf{P} = \mathbf{I} - \mathbf{Z}_1(\mathbf{Z}_1'\mathbf{Z}_1)\mathbf{Z}_1'$ permits us to define $\mathbf{Z}^* = \mathbf{P}\mathbf{Z}_2$. In the periodic case, $\mathbf{Z}_1$ has columns $(1,\ldots,1)$ and $(\bar{x}_1,\ldots,\bar{x}_N)$, where $\bar{x}_i = \int_{\mathcal{T}} x_i(s)\,ds$ for each i. Thus $\mathbf{P}$ is the $N \times N$ matrix that projects any N-vector z to its residuals from its linear regression on $\bar{x}_i$.

Continuing with this partitioning process, let ζ_1 be the vector of the first $M_0 + 1$ components of ζ, and let ζ_2 be the remaining components of ζ. Then, by multiplying both sides by $\mathbf{Z}'$ the constraint

$$\mathbf{Z}\zeta = \mathbf{Z}_1\zeta_1 + \mathbf{Z}_2\zeta_2 = \hat{y}$$

implies that

$$\mathbf{Z}_1'\mathbf{Z}_1\zeta_1 + \mathbf{Z}_1'\mathbf{Z}_2\zeta_2 = \mathbf{Z}_1'\hat{y}. \tag{10.21}$$

Solving for ζ_1 alone gives

$$\zeta_1 = (\mathbf{Z}_1'\mathbf{Z}_1)^{-1}\mathbf{Z}_1'(\hat{y} - \mathbf{Z}_2\zeta_2) \text{ and } \mathbf{Z}_1\zeta_1 = \mathbf{P}(\hat{y} - \mathbf{Z}_2\zeta_2). \tag{10.22}$$

In the periodic case, equation (10.22) indicates that ζ_1 is obtained by linear regression of the values $(\hat{y} - \mathbf{Z}_2\beta_2)$ on the vector with components $\bar{x}_i$. Thus, once ζ_2 has been determined, we can find ζ_1.

Now substitute the solution (10.22) for ζ_1 into the constraint (10.21) and rearrange to show that we can find ζ_2 by solving the minimization problem

$$\min_{\zeta_2}\{\zeta_2'\mathbf{K}_2\zeta_2\} \text{ subject to } \mathbf{Z}^*\zeta = \mathbf{P}\hat{y} \tag{10.23}$$

using the fact that $\zeta'\mathbf{K}\zeta = \zeta_2'\mathbf{K}_2\zeta_2$.

Let $\mathbf{R}$ be defined as the Moore-Penrose g-inverse

$$\mathbf{R} = (\mathbf{Z}^*\mathbf{K}_2^{-1}\mathbf{Z}^{*\prime})^+. \tag{10.24}$$

The solution of the minimization (10.23) is then given by

$$\zeta_2 = \mathbf{K}_2^{-1}\mathbf{Z}^{*\prime}\mathbf{R}\hat{y} \tag{10.25}$$

and the minimum value of the objective function $\zeta'\mathbf{R}\zeta$ is therefore given by

$$\begin{aligned} \zeta'\mathbf{R}\zeta &= \zeta_2'\mathbf{K}_2\zeta_2 = \hat{y}'\mathbf{R}\mathbf{Z}^{*\prime}\mathbf{K}_2^{-1}\mathbf{K}_2\mathbf{K}_2^{-1}\mathbf{Z}^*\mathbf{R}\hat{y} \\ &= \hat{y}'\mathbf{R}\mathbf{R}^+\mathbf{R}\hat{y} = \hat{y}'\mathbf{R}\hat{y}. \end{aligned} \tag{10.26}$$

This is the assumption we made above in defining the two-step procedure, and moreover we have now defined the matrix $\mathbf{R}$.

We can sum up this discussion by setting out the following algorithm for functional interpolation:

Step 1: Calculate matrices $\mathbf{P} = \mathbf{I} - \mathbf{Z}_1(\mathbf{Z}_1'\mathbf{Z}_1)\mathbf{Z}_1'$ and $\mathbf{Z}^* = \mathbf{P}\mathbf{Z}_2$. In effect, the columns of $\mathbf{Z}^*$ are the residuals from a standard regression of the corresponding columns of $\mathbf{Z}_2$ on the design matrix $\mathbf{Z}_1$.

Step 2: Compute $\mathbf{R}$ as defined in (10.24) above.

Step 3: Compute ζ_2 from (10.24) and use (10.22) to find ζ_1.

Of course, if all we require is the roughness of ζ, then we can find $\hat{y}'\mathbf{R}\hat{y}$ from (10.24) without actually calculating ζ.

Finally, returning to our two-stage technique for smoothing, we can carry out the first step by solving the equation

$$(\mathbf{I} + \lambda\mathbf{R})\hat{y} = y.$$

Note that if $\mathbf{R}$ is either diagonal (as for the Fourier basis) or band-structured (as for the B-spline basis) the solution to this equation is very cheap to compute, and hence trying out various values for λ is quite feasible.

If we are dealing with a large data set by truncating or restricting the basis expansion to a reasonable dimensionality K as described in Section 10.6.3, then we wish in general only to assess the roughness of variates of the form $\mathbf{Z}\zeta$ for known ζ with $\zeta_j = 0$ for $j > m$. It is usually more appropriate to calculate $\zeta'\mathbf{R}\zeta$ directly for such variates directly if necessary.

10.8 Cross-validation and regression diagnostics

We have already noted the possibility of choosing the smoothing parameter λ by cross-validation. Various economies are possible in calculating the cross-validation score $\mathrm{CV}(\lambda)$ as defined in (10.11).

Let $\mathbf{S}$ be the so-called *hat matrix* of the smoothing procedure which maps the data values y to their fitted values $\hat{y}$ for any particular value of λ. A calculation described, for example, in Section 3.2 of Green and Silverman (1994), shows that the cross-validation score satisfies

$$\mathrm{CV}(\lambda) = \sum_{i=1}^{N} \left(\frac{y - \hat{y}_i}{1 - S_{ii}} \right)^2 .$$

If N is of moderate size and we are using the algorithm described in Section 10.7, then the hat matrix satisfies

$$\mathbf{S} = (\mathbf{I} + \lambda \mathbf{R})^{-1}.$$

The calculation (10.24) of the symmetric matrix $\mathbf{R}$ as a Moore-Penrose g-inverse means that we have expressed $\mathbf{R}$ as $\mathbf{U}\,\mathrm{diag}(\rho_1, \rho_2, \ldots)\mathbf{U}'$, where $\mathbf{U}$ is an orthogonal matrix and the ρ_i are the eigenvalues of $\mathbf{R}$. It follows that the diagonal elements of $\mathbf{S}$ are given by

$$S_{ii} = \sum_j (1 + \lambda \rho_j)^{-1} u_{ij}^2.$$

If N is large and we are considering an expansion in a moderate number K of basis functions, as in Section 10.6.3, then we can find the diagonal elements of $\mathbf{S}$ directly from

$$\mathbf{S} = \mathbf{Z}(\mathbf{Z}'\mathbf{Z} + \lambda \mathbf{R})^{-1}\mathbf{Z}'.$$

From $\mathbf{S}$, we can also compute an indicator of the effective degrees of freedom used up in the approximation. Either $\mathrm{trace}\,\mathbf{S}$ or $\mathrm{trace}\,\mathbf{S}^2$ were recommended for this purpose by Buja, Hastie, and Tibshirani (1989). For the fit in Figure 10.5, defined by minimizing the cross-validation criterion, the effective degrees of freedom are estimated to be $\mathrm{trace}\,\mathbf{S} = 6.8$.

Another important use of the hat matrix $\mathbf{S}$ is in constructing various regression diagnostics. The diagonal elements of the hat matrix are often called *leverage values*; they determine the amount by which the fitted value $\hat{y}_i$ is influenced by the particular observation y_i. If the leverage value is particularly high, the fitted value needs to be treated with some care. Two standard ways of assessing the regression fit are to examine the raw residuals $y_i - \hat{y}_i$ and the *deleted residuals* $(y_i - \hat{y}_i)/(1 - S_{ii})$; the latter give the residual between y_i and the value predicted from the data set with case i deleted. We refer readers to works on regression diagnostics such as Cook and Weisberg (1982).

Figure 10.8 shows a plot of deleted residuals against fitted values for the log precipitation and temperature example, with the smoothing parameter chosen by cross-validation. The three observations with small predicted values have somewhat larger leverage values (around 0.4) than the others (generally in the range 0.1 to 0.2). This is not surprising, given that they are somewhat isolated from the main part of the data.

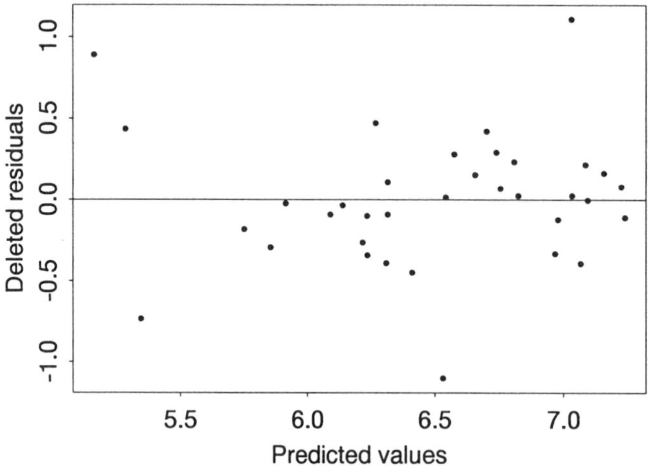

FIGURE 10.8. Deleted residuals from the fitted prediction of log annual precipitation from overall temperature pattern.

10.9 Functional regression and integral equations

Functional interpolation and regression can be viewed as a different formalization of a problem already considered in detail elsewhere, reconstructing a curve given certain indirect observations. Suppose that g is a curve of interest, and that we have noisy observations of a number of linear functionals $l_i(g)$. Such a problem was explored by Engle, Granger, Rice and Weiss (1986); see also Section 4.7 of Green and Silverman (1994). The problem involved reconstructing the effect of temperature t on electricity consumption, so that $g(t)$ is the expected use of electricity per consumer on a day with average temperature t. Various covariates were also considered, but these need not concern us here.

Electricity bills are issued on various days and always cover the previous 28 days. For bills issued on day i, the average consumption (after correcting for covariates) would be modelled to satisfy

$$\frac{1}{28}\mathrm{E}Y_i = \langle \theta_i, g \rangle$$

where θ_i is the probability density function of temperature over the previous 28 day period. By setting $x_i = 28\theta_i$ and $\beta = g$, we see that this problem falls precisely into the functional regression context, and indeed the method used by the original authors to solve it corresponds precisely to the regularization method we have set out.

More generally, regularization is a very well known tool for solving integral equations; see, for example, Section 12.3 of Delves and Mohamed (1985).

11

Functional linear models for functional responses

11.1 Introduction

The aim of Chapter 10 was to predict a scalar response y from a functional covariate x. We now consider a fully functional linear model in which both the response y and the covariate x are functions. For instance, in the Canadian weather example, we might wish to investigate to what extent we can predict the complete log precipitation profile `LPrec` of a weather station from information in its complete temperature profile `Temp`.

We assume that we have N observed functions y_i, each with associated covariate function x_i. The x_i are functions on an interval $\mathcal{T}_X$ and the y_i on an interval $\mathcal{T}_Y$. There is no need for the intervals to be the same. In this chapter, we use the notation $\mathbf{x}$ and $\mathbf{y}$ for the vectors of functions with components being the functions x_i and y_i.

When y_i is a function rather than a multivariate observation, the usual linear model becomes

$$\hat{y}_i(t) = \alpha(t) + \int_{\mathcal{T}_X} x_i(s)\beta(s,t)\,ds. \tag{11.1}$$

Clearly, the regression function β must be a function of t as well as of s, but the principle of the linear effect of the covariates x_i on the parameter function β remains the same. We can interpret the regression function β for a fixed value of t, denoted by $\beta(\cdot, t)$, as the relative weights placed on temperature for all possible times to predict log precipitation at time t.

The function α plays the part of the constant term in the standard regression setup. One simple approach to the estimation of α is to centre the observed y_i and the given x_i by subtracting their sample average functions $\bar{y}$ and $\bar{x}$ and to consider the model as

$$\hat{y}_i(t) - \bar{y}(t) = \int_{\mathcal{T}_X} x_i^*(s)\beta(s,t)\,ds \tag{11.2}$$

where $x_i^* = x_i - \bar{x}$. This constrains the value of $\alpha(t)$ to be equal to $\bar{y}(t) - \int \bar{x}(s)\beta(s,t)\,ds$. Of course, on occasion we may wish to estimate α in some other way, for example by incorporating some smoothing, but we do not consider this possibility in any detail.

In whatever way we deal with α, the fitting criterion in the present context is the extension of the residual sum of squares LMSSE as defined in (10.2) to assess lack of fit across the index t as well as i. A simple measure that combines information across t in $\mathcal{T}_Y$ is the integrated squared residual

$$\text{LMISE} = \|\mathbf{y} - \hat{\mathbf{y}}\|^2 = \sum_{i=1}^{N} \int_{\mathcal{T}_Y} [y_i(t) - \alpha(t) - \int_{\mathcal{T}_X} x_i(s)\beta(s,t)\,ds]^2\,dt. \tag{11.3}$$

11.2 Estimating β by basis representations

In this section we obtain an expression for LMISE using basis functions that can be used to fit the model (11.1) in practice. For the moment, we use the special case (11.2) of the model, and centre the x_i and y_i, writing $y_i^* = y_i - \bar{y}$. Expand the x_i^* in a basis ϕ_j and the y_i^* in a basis ψ_k, to give

$$x_i^* = \sum_{j=1}^{J} c_{ij}\phi_j = c_i'\phi$$

and

$$y_i^* = \sum_{k=1}^{K} d_{ik}\psi_k = d_i'\psi,$$

where ϕ and ψ are the J- and K-vectors of the respective basis functions. We denote the matrices of coefficients by $\mathbf{C}$ and $\mathbf{D}$, so that we can write these expressions in the function form

$$\mathbf{x}^* = \mathbf{C}\phi$$

and

$$\mathbf{y}^* = \mathbf{D}\psi.$$

We consider the expression of β as a double expansion

$$\beta(s,t) = \sum_{j=1}^{J} \sum_{k=1}^{K} b_{jk}\phi_j(s)\psi_k(t) = \phi(s)'\mathbf{B}\psi(t), \tag{11.4}$$

where $\mathbf{B}$ is a $J \times K$ matrix of coefficients b_{jk}, or, more compactly, as $\beta = \phi'\mathbf{B}\psi$. Define $\mathbf{J}_\phi$ and $\mathbf{J}_\psi$ to be the matrices of inner products between the elements of the ϕ and ψ bases, respectively. Thus,

$$\mathbf{J}_\phi = \int_{\mathcal{T}_X} \phi(s)\phi(s)'\,ds$$

and

$$\mathbf{J}_\psi = \int_{\mathcal{T}_Y} \psi(s)\psi(s)'\,ds.$$

Substitute the basis expansions of x_i and β into (11.1) to give

$$\hat{\mathbf{y}}^*(t) = \int \mathbf{C}\phi(s)\phi(s)'\mathbf{B}\psi(t)\,ds = \mathbf{C}\mathbf{J}_\phi\mathbf{B}\psi(t).$$

If we let $\hat{\mathbf{D}}$ be the matrix of coefficients of the basis expansion of the vector of predictors $\hat{\mathbf{y}}^*$ (corresponding to the matrix $\mathbf{D}$ for the vector $\mathbf{y}^*$), we obtain the matrix form of the model

$$\hat{\mathbf{D}} = \mathbf{C}\mathbf{J}_\phi\mathbf{B}.$$

Now we can get an expression for the integrated squared residual:

$$\|\hat{y}_i - y_i\|^2 = \|\hat{y}_i^* - y_i^*\|^2 = \{(\hat{\mathbf{D}} - \mathbf{D})\mathbf{J}_\psi(\hat{\mathbf{D}} - \mathbf{D})'\}_{ii}$$

and, finally,

$$\texttt{LMISE}(\mathbf{B}) = \text{trace}\{(\mathbf{C}\mathbf{J}_\phi\mathbf{B} - \mathbf{D})\mathbf{J}_\psi(\mathbf{C}\mathbf{J}_\phi\mathbf{B} - \mathbf{D})'\}, \tag{11.5}$$

a quadratic form in the unknown coefficient matrix $\mathbf{B}$. If the bases ϕ and ψ are orthonormal, then of course the matrices $\mathbf{J}_\phi$ and $\mathbf{J}_\psi$ are the identity matrices of order J and K, respectively, and the expression for LMISE($\mathbf{B}$) simplifies accordingly.

11.3 Fitting the model by basis expansions

11.3.1 Some linear algebra preliminaries

First of all let us consider the minimization of the quantity LMISE($\mathbf{B}$), as given in (11.5). In the case where $\mathbf{J}_\phi$ and $\mathbf{J}_\psi$ are identity matrices, the

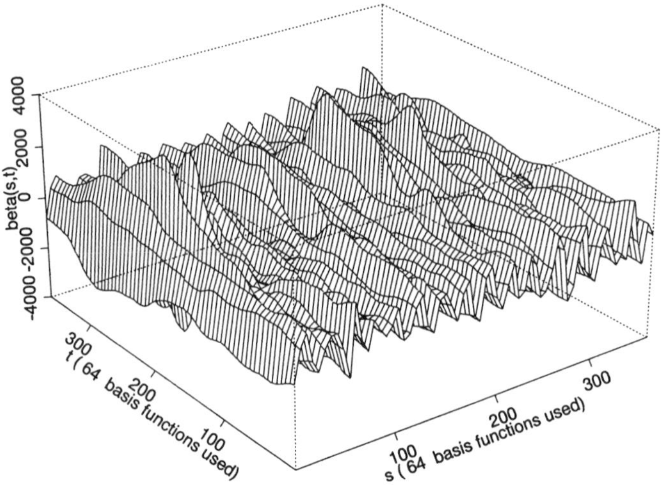

FIGURE 11.1. The functional parameter function β for the prediction of log precipitation from temperature, estimated direct from the data. The value $\beta(s,t)$ shows the influence of temperature at time s on log precipitation at time t.

matrix **B** will minimize (11.5) if and only if

$$\mathbf{C}'\mathbf{C}\mathbf{B} = \mathbf{C}'\mathbf{D}$$

so that

$$\mathbf{B} = (\mathbf{C}'\mathbf{C})^{-}\mathbf{C}'\mathbf{D}. \tag{11.6}$$

The matrix **B** is easily found by using the SVD of **C**. Write $\mathbf{C} = \mathbf{U}\Delta_C\mathbf{V}'$ where Δ_C is a diagonal matrix with strictly positive diagonal elements and **U** and **V** have orthonormal columns. Then $\mathbf{C}'\mathbf{C} = \mathbf{V}\Delta_C^2\mathbf{V}'$, and hence the Moore-Penrose g-inverse of $\mathbf{C}'\mathbf{C}$ is $\mathbf{V}\Delta_C^{-2}\mathbf{V}'$. Substituting into (11.6) gives

$$\mathbf{B} = \mathbf{V}\Delta_C^{-1}\mathbf{U}'\mathbf{D}. \tag{11.7}$$

In the example discussed in detail below, we use Fourier bases which are orthonormal, and so we need not consider the case of more general $\mathbf{J}_\phi$ and $\mathbf{J}_\psi$. For the more general case, the results corresponding to (11.6) and (11.7) are notationally more complicated but are easily stated. We require that

$$\mathbf{J}_\phi\mathbf{C}'\mathbf{C}\mathbf{J}_\phi\mathbf{B}\mathbf{J}_\psi = \mathbf{J}_\phi\mathbf{C}'\mathbf{D}\mathbf{J}_\psi.$$

Provided $\mathbf{J}_\psi$ is nonsingular (i.e. the functions ψ are not linearly dependent), this implies that

$$\mathbf{J}_\phi \mathbf{C}'\mathbf{C}\mathbf{J}_\phi \mathbf{B} = \mathbf{J}_\phi \mathbf{C}'\mathbf{D}$$

so that $\mathbf{C}$ in (11.6) is replaced by $\mathbf{C}\mathbf{J}_\phi$, and $\mathbf{J}_\psi$ plays no role in defining the coefficient matrix $\mathbf{B}$.

11.3.2 Fitting the model without regularization

We now apply this methodology to the data on Canadian climate, considering the detailed data that give daily precipitation and temperature. Because all the functions in this example are intrinsically periodic, we can expand both the log precipitations and the temperatures in Fourier series. We preprocessed the data by fitting a Fourier series with 64 terms, applying a roughness penalty smoother by tapering the series to eliminate very local variation. Therefore $\mathbf{C}$ and $\mathbf{D}$ are 64×35 matrices, and $\mathbf{B}$ as found by (11.7) is a 64×64 matrix. Substituting into (11.4) gives the estimated function β plotted in Figure 11.1.

We see that the function β estimated by this method is extremely variable. It is also the case that this β gives perfect prediction of the given data. Although this is superficially attractive, it does not make physical sense: whatever influence temperature patterns may have on precipitation patterns, it is naïve to imagine that the precipitation pattern at a place can be entirely accounted for by its temperature pattern.

The reason for this overfitting is an extension of the discussion in Section 10.2 on functional interpolation. Let $\mathtt{Temp}$ be the vector of observed covariate functions and $\mathtt{LPrec}$ the vector of observed log precipitations. Consider any fixed t. As in Section 10.2, we can find a number α_t and a function β_t such that, for all i,

$$\mathtt{LPrec}_i(t) = \alpha_t + \langle \mathtt{Temp}_i, \beta_t \rangle$$

without any error. Writing $\alpha(t)$ for α_t, and $\beta(s,t)$ for β_t yields

$$\mathtt{LPrec}_i(t) = \alpha(t) + \int_{\mathcal{T}_X} \mathtt{Temp}_i(s)\beta(s,t)\, ds,$$

a perfect fit to the observed data. Just as in Chapter 10, we must regularize the functional predictor variable. This is discussed in the next section.

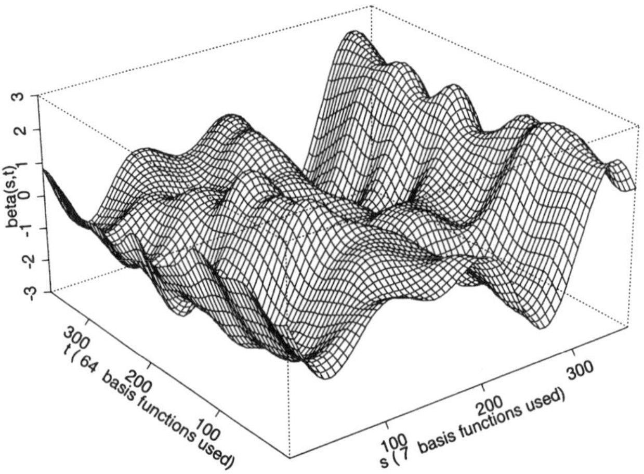

FIGURE 11.2. Perspective plot of estimated β function truncating the basis for the temperature covariates to 7 terms.

11.4 Regularizing the fit

11.4.1 Restricting the basis of the predictor variable

Just as in the case of the prediction of a scalar response from a functional predictor, a natural approach is to truncate the basis ϕ in which the predictors are expressed. Instead of expanding the x_i^* in a series of length J, we choose some $J_0 < J$ and discard all but the first J_0 terms in the expansion of the x_i^*. Then $\mathbf{C}$ is an $n \times J_0$ matrix, and $\mathbf{B}$ is $J_0 \times K$. Otherwise, the numerical details are the same; there is considerable computational economy because the complexity of the SVD and of other matrix calculations is lessened by the reduction in the sizes of the matrices involved.

We apply this approach to the weather data, setting $J_0 = 7$. Figure 11.2 shows the resulting estimated β function. The resulting prediction of the annual *pattern* of log precipitation at four selected stations is demonstrated in Figure 11.3. In this figure, both the original data and the predictions for log precipitation have their annual mean subtracted, to highlight the pattern of precipitation rather than its overall level. The precipitation pattern is quite well predicted except for Edmonton, which therefore has a precipitation pattern different from other weather stations with similar temperature profiles.

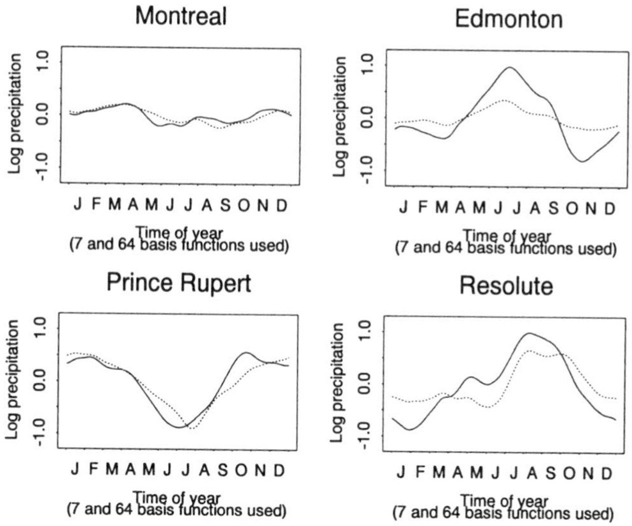

FIGURE 11.3. Original data (solid) and predictions (dashed) of log precipitation relative to annual mean for each of 4 weather stations. The prediction is carried out using an estimated β function with the temperature covariate truncated to 7 terms.

Although the plot of the estimated β function demonstrates a more plausible influence of temperature pattern on precipitation pattern, it is not easy to interpret. As a function of t for any fixed s it is irregular, and this irregularity is easily explained. Because every Fourier coefficient of log precipitation is allowed to be predicted by the temperature covariate, the prediction contains frequency elements at all levels. By the arguments given in Chapter 10, we expect that each individual Fourier coefficient will be predicted sensibly as a scalar response. However, in putting these together to give a functional prediction, the high-frequency terms are given inappropriate weight. From a common-sense point of view, we cannot expect overall temperature patterns to affect a very high frequency aspect of log precipitation at all. To address this difficulty, we consider the idea of restricting or truncating the ψ basis in terms of which the functional response variable is expanded.

11.4.2 Restricting the basis of the response variable

In this section, we consider the approach of truncating the ψ basis, allowing the prediction of only low-frequency aspects of the response variable. In our example, this would correspond to the idea that

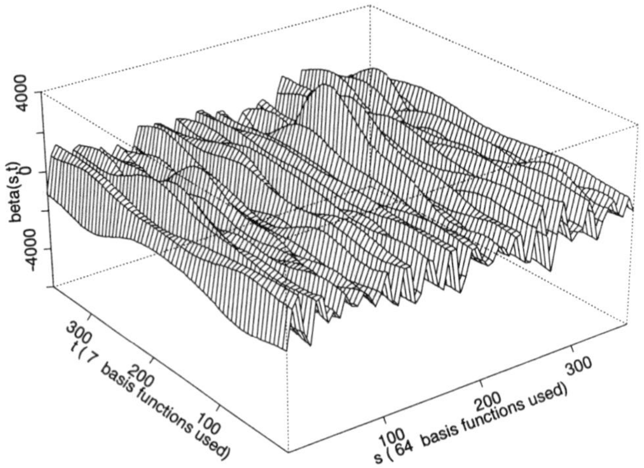

FIGURE 11.4. Perspective plot of estimated β function truncating the basis for the log precipitations to 7 terms.

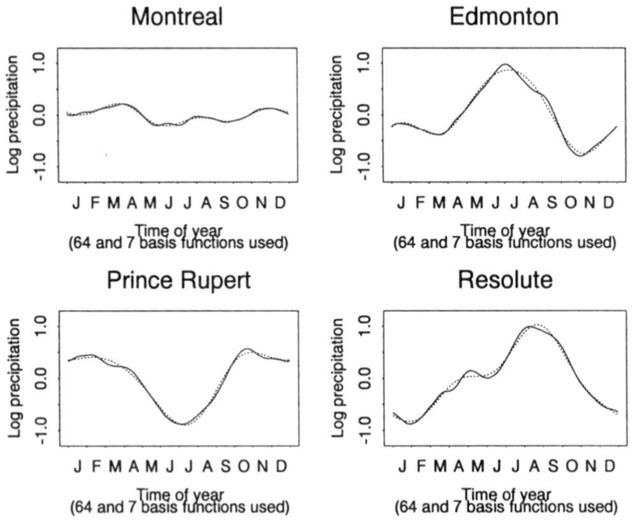

FIGURE 11.5. Original data (solid) and predictions (dashed) of log precipitation relative to annual mean for each of four weather stations. The prediction is carried out using an estimated β function with the basis for the log precipitations truncated to 7 terms.

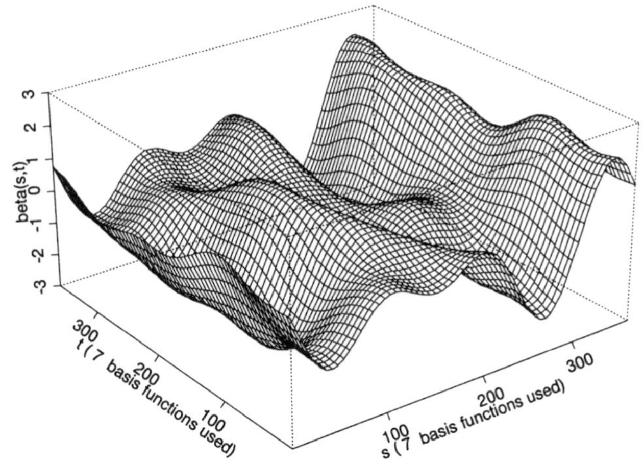

FIGURE 11.6. Perspective plot of estimated β function truncating both bases to seven terms.

the very fine detail of log precipitation could not be predicted from temperature. For the moment, suppose that we do not truncate the ϕ basis, but that we allow only $K_0 = 7$ terms in the expansion of the y_i^*, with corresponding adjustments to the matrices $\mathbf{D}$ and $\mathbf{B}$. Figures 11.4 and 11.5 show the resulting β functions and sample predictions. The predictions are smooth, but otherwise very close to the original data. The function β is similar in overall character to the unsmoothed function shown in Figure 11.1, except that it is smoother as a function of t. However, it is excessively rough as a function of s. Thus, although the predictions are aesthetically attractive as smooth functions, they provide an optimistic assessment of the quality of the prediction, and an implausible mechanism by which the prediction takes place.

11.4.3 Restricting both bases

Sections 11.4.1 and 11.4.2 illustrated advantages in truncating both the ϕ basis of the predictors and the ψ basis of the responses to obtain useful and sensible estimates. It should be stressed that the reason for doing this is not the same in both cases. Truncating the ϕ basis for the covariates is essential to avoid over-fitting, while the ψ basis is truncated to ensure that the predictions are smooth.

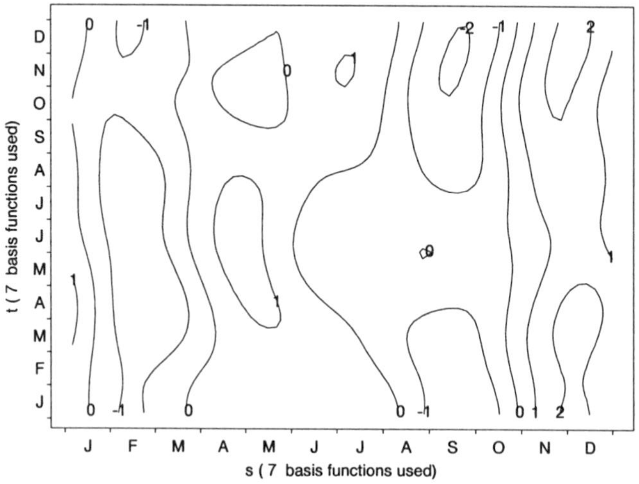

FIGURE 11.7. Contour plot of estimated β function truncating both bases to seven terms.

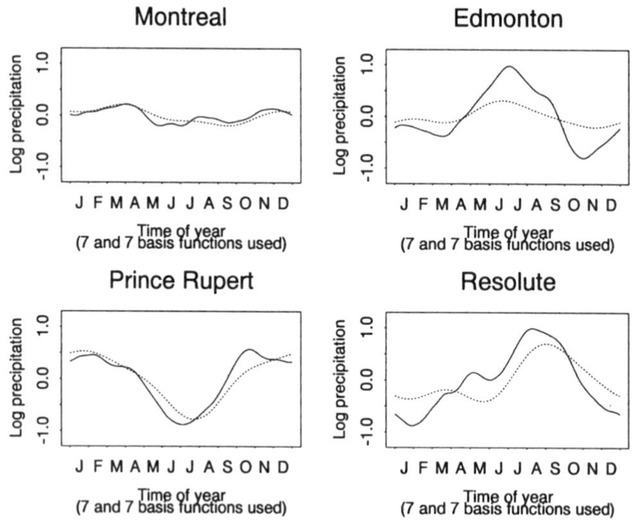

FIGURE 11.8. Original data (solid) and predictions (dashed) of log precipitation relative to annual mean for each of four weather stations. The prediction is carried out using an estimated β function with the both bases truncated to seven terms.

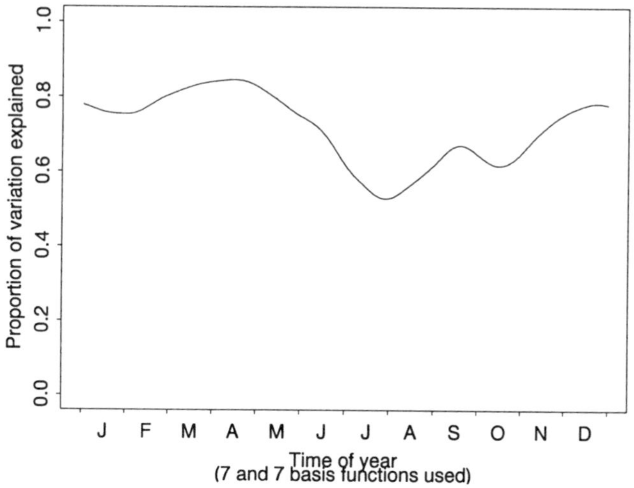

FIGURE 11.9. Proportion of variance of log precipitation explained by a linear model based on daily temperature records. The prediction is carried out using an estimated β function with both bases truncated to seven terms.

Let us combine these different reasons for truncating the bases, and truncate both the predictor basis ϕ and the response basis ψ. Figures 11.6, 11.7 and 11.8 show the effects of truncating both bases to seven terms. We can discern several aspects of the effect of temperature on log precipitation. Temperature in February is negatively associated with precipitation throughout the year. Temperature around May is positively associated with precipitation in the summer months. Temperature in September has a strong negative association with precipitation in the autumn and winter, and finally, temperature in December associates positively with precipitation throughout the year, particularly with winter precipitation.

11.5 Assessing goodness of fit

There are various ways of assessing the fit of a functional linear model as estimated in Section 11.4. An approach borrowed from the conventional linear model is to consider the squared correlation function

$$R^2(t) = 1 - \sum_i \{\hat{y}_i(t) - y_i(t)\}^2 \Big/ \sum_i \{y_i(t) - \bar{y}(t)\}^2.$$

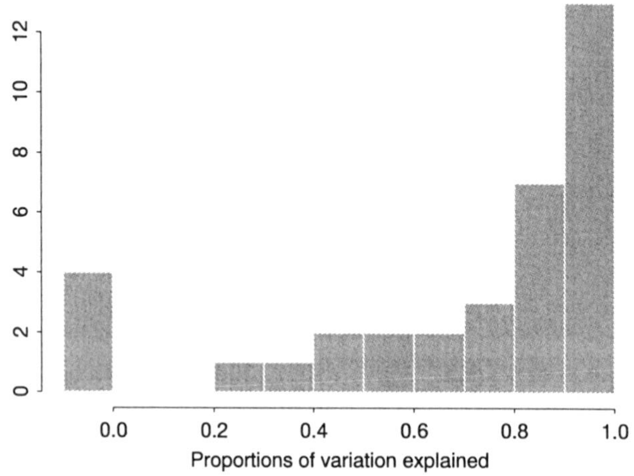

FIGURE 11.10. Histogram of individual proportions of variance R_i^2 in log precipitation explained by a linear model based on daily temperature records. The prediction is carried out using an estimated β function with both bases truncated to seven terms. The left-hand cell of the histogram includes all cases with negative R_i^2 values.

If we require a single numerical measure of fit, then the average of R^2 over t is useful, but using the entire function R^2 offers more detailed information about the fit. Figure 11.9 plots the R^2 function for the fit to the log precipitation data in Figure 11.6. The fit is generally reasonable, and is particularly good in the first five months of the year.

A complementary approach to goodness of fit is to consider an overall R^2 measure for each individual functional datum, defined by

$$R_i^2 = 1 - \int \{\hat{y}_i(t) - y_i(t)\}^2 \, dt \Big/ \int \{y_i(t) - \bar{y}(t)\}^2 \, dt.$$

For the four particular stations plotted in Figure 11.8, for instance, the values of R_i^2 are 0.96, 0.67, 0.63 and 0.81 respectively, illustrating that Montreal and Resolute are places whose precipitations fit closely to those predicted by the model on the basis of their observed temperature profiles; for Edmonton and Prince Rupert the fit is of course still quite good in that the temperature pattern accounts for over 60% of the variation of the log precipitation from the overall population mean. However, Figure 11.8 demonstrates that the pattern of precipitation, judged by comparing the predictions with the original data after subtracting the annual mean for the individual places, is

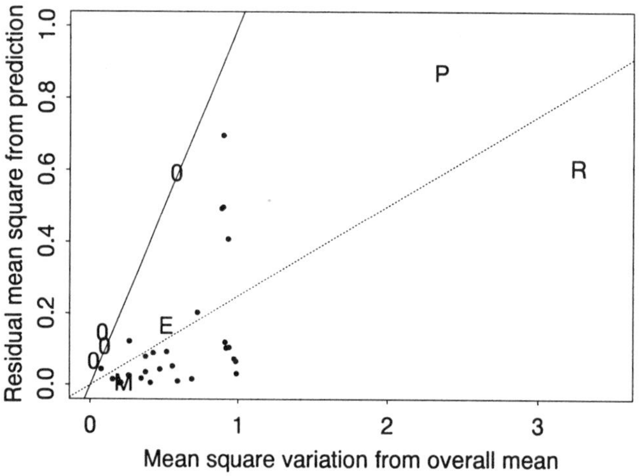

FIGURE 11.11. Comparison, for log precipitation, between mean square prediction errors and mean square variation from overall mean of log precipitation. The prediction is carried out using an estimated β function with both bases truncated to seven terms. The points for Montreal, Edmonton, Prince Rupert and Resolute are marked as M, E, P and R respectively. The points marked 0 yield negative R_i^2 values. The lines $y = x$ and $y = 0.25x$ are drawn on the plot as solid and dotted, respectively.

predicted only moderately well for Resolute and is not well predicted for Edmonton. Figure 11.10 displays a histogram of all 35 R_i^2 values. At most of the stations, the R_i^2 value indicates reasonable or excellent prediction, but for a small proportion the precipitation pattern is not at all well predicted. Indeed, four stations (Dawson, Schefferville, Toronto and Prince George) have negative R_i^2 values, indicating that for these places the population mean $\bar{y}$ actually gives a better fit to y_i than does the predictor $\hat{y}_i$.

To investigate this effect further, we use Figure 11.11 to show a plot of the residual mean square prediction error $\int(\hat{y}_i - y_i)^2$ against the mean square variation from the overall mean, $\int(y_i - \bar{y})^2$. The four places with negative values of R_i^2 are indicated by 0's on the plot. Each of the four places plotted in Figure 11.8 is indicated by the initial letter of its name. For most places the predictor has about one quarter the mean square error of the overall population mean, and for many places the predictor is even better. The four places that yielded a negative value of R_i^2 did so because they were close (in three cases very close) to the overall population mean, not because the predictor did not work well for them. To judge accuracy of prediction for an individual place, it is clear that one needs to look a little further than just at the statistic R_i^2.

It is possible to conceive of an F-ratio function for the fit. Substituting into equation (11.2), say, we have

$$\hat{y}_i(t) - \bar{y}(t) = \sum_{j=1}^{J_0} C_{ij} \left(\sum_{k=1}^{K_0} B_{jk} \psi_k(t) \right) = \sum_{j=1}^{J_0} C_{ij} \theta_j(t).$$

By analogy with the standard linear model, we can ascribe $K_0 - 1$ degrees of freedom to the pointwise sum of squares $\sum_i \{\hat{y}_i(t) - \bar{y}(t)\}^2$ and $n - K_0$ degrees of freedom to the residual sum of squares $\sum_i \{y_i(t) - \hat{y}_i(t)\}^2$. An F-ratio plot would be constructed by plotting

$$\mathtt{FRATIO}(t) = \frac{\sum_i \{\hat{y}_i(t) - \bar{y}(t)\}^2 / (K_0 - 1)}{\sum_i \{y_i(t) - \hat{y}_i(t)\}^2 / (n - K_0)}.$$

However, the parameters $\theta_j(t)$ are not directly chosen to give the best fit of $\hat{y}_i(t)$ to the observed $y_i(t)$, and so the classical distribution theory of the F-ratio could be used only as an approximation to the distribution of $\mathtt{FRATIO}(t)$ for each t.

Figure 11.12 plots the F-ratio for the fit to the log precipitation data. The upper 5% and 1% points of the $F_{6:28}$ distribution are given; within this model, this indicates that the effect of daily temperature on precipitation is highly significant overall.

We have not given much attention to the method by which the truncation parameters J_0 and K_0 could be chosen in practice. For many smoothing and regularization problems, the appropriate method of choice is probably subjective. The different roles of J_0 and K_0 lead to different ways of considering their automatic choice, if one is desired. The variable J_0 corresponds to a number of terms in a regression model, and so we could use a variable selection technique from conventional regression, possibly adapted to give a functional rather than a numerical criterion, to indicate a possible value. On the other hand, K_0 is more akin to a smoothing parameter in a smoothing method, and so a method such as cross-validation might be a more appropriate choice. These questions are interesting topics for future investigation and research.

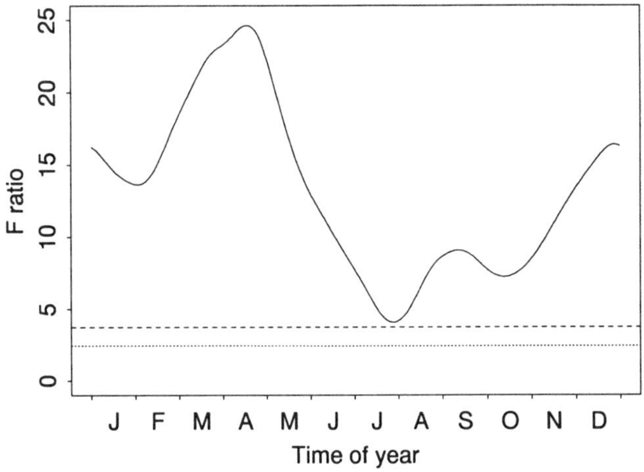

FIGURE 11.12. A plot of the F-ratio function for the prediction of log precipitation from daily temperature data. The prediction is carried out using an estimated β function with both bases truncated to seven terms. The horizontal lines show the upper 5% and 1% points of the $F_{6:28}$ distribution.

11.6 Bivariate roughness penalties for β

Just as in other smoothing contexts, an obvious alternative to placing restrictions on β by truncating the basis expansion is to define a roughness penalty $\text{PEN}(\beta)$ of some kind. We would then minimize $\text{LMISE}(\beta) + \lambda \times \text{PEN}(\beta)$, where $\text{LMISE}(\beta) = \text{LMISE}$ as defined in (11.1), and $\alpha(t) = \bar{y}(t) - \int \bar{x}(s)\beta(s,t)\,ds$ as in Section 11.1. As usual, λ is a smoothing parameter. The ideas of this section are very much tentative suggestions for future research, and we provide only a brief discussion of some possibilities.

A useful roughness penalty can be constructed by finding the integral of the square of some differential operator acting on β. One possibility is to use a roughness penalty such as the thin plate spline penalty

$$\text{PEN}(\beta) = \iint \{(D_s^2 + 2D_sD_t + D_t^2)\beta(s,t)\}^2\,ds\,dt \qquad (11.8)$$

discussed in Chapter 7 of Green and Silverman (1994), for example. This penalty has the property that the only functions with roughness zero are linear in s and t. It also has the property that the penalty is isotropic: If the coordinates are rotated, the penalty is unaffected.

We can impose periodic boundary conditions, such as those appropriate for the Canadian climate example, by restricting attention to functions whose values and derivatives obey the appropriate periodic continuity conditions; we then consider the integrals in (11.8) evaluated on the periodic version of the interval on which s and t lie. In this case, the only functions with zero roughness are constant.

Because the arguments s and t play different roles, there is no particular need to use an isotropic roughness penalty such as (11.8). Another possibility is to use a penalty such as $\int\int \{(D_s + D_t)\beta\}^2$, which is zero if β is a function of $s - t$ alone, corresponding to stationary dependence of y on x.

Probably the most straightforward, if not necessarily the most economical, computational approach is to express any quadratic roughness penalty as a quadratic form in the elements of **B**, where **B** is the matrix of coefficients of the expansion (11.4) of β in terms of basis functions. The penalized integrated squared residual $\text{LMISE}(\beta) + \lambda \times \text{PEN}(\beta)$ is then itself a quadratic form in **B**, and can therefore be minimized to find the estimate. The practical details, and the numerical conditioning, depend somewhat on the particular bases chosen. Another possible approach to the definition of a roughness penalty is to construct the quadratic function of **B** directly, and this would provide a more immediate generalization of the truncated basis function method.

It is possible to set out a cross-validation approach to the selection of the smoothing parameter λ, based on repredicting each y_i by the action of the covariate x_i on the function β estimated from all the other data. We obtain the cross-validation score by summing the integrated squared prediction errors over all i, as usual. We can use a similar approach within the basis truncation method in Section 11.4.

11.7 Other linear models for functional data

The functional linear models introduced in Chapters 9 and 10 and in this chapter by no means exhaust the possibilities. In this section we briefly review a few more that can be interesting in applications.

11.7.1 More complicated covariate structures

In this chapter, we have concentrated for simplicity on the case where the covariate function (in our example the temperature) is a single real-valued function. In principle, it is easy to contemplate extending the ideas to problems where there are vector covariate functions and/or multivariate covariates. For example, if there were a pair of covariate functions (x_i, z_i) and a vector of covariates $\mathbf{w}_i$ for each observed function y_i, the linear model would become

$$\hat{y}_i(t) = \alpha(t) + \int x_i(s)\beta_1(s,t)\,ds + \int z_i(v)\beta_2(v,t)\,dv + \mathbf{w}_i'\beta_3(t),$$

where β_1 and β_2) were suitable bivariate functions and β_3 is a vector of functions. We can expand the parameter functions β in terms of suitable bases, and basis truncation and roughness penalty ideas are simply extended. Again, the full details of this approach need to be worked out in the context of any particular application, and we shall not discuss it in any detail.

11.7.2 The pointwise multivariate model

An extension of the linear models considered in Chapter 9 is

$$y_i(t) = \sum_j^q z_{ij}(t)\beta_j(t) + \epsilon_i(t). \tag{11.9}$$

In vector/matrix form this model may be written

$$\mathbf{y}(t) = \mathbf{Z}(t)\beta(t) + \epsilon(t)$$

where the matrix $\mathbf{Z}(t)$ contains the entries $z_{ij}(t)$, and the functional vectors $\mathbf{y}$, β and ϵ have the appropriate definitions.

This can be called a pointwise multivariate model, because when the argument t is fixed, the problem of estimating vector $\beta(t)$ is the familiar one of ordinary least squares. The solution becomes

$$\hat{\beta}(t) = [\mathbf{Z}(t)'\mathbf{Z}(t)]^{-1}\mathbf{Z}(t)'\mathbf{y}(t). \tag{11.10}$$

In practice we would evaluate $\hat{\beta}$ only at a finite number of points t, and some form of interpolation would be used between these points. It may also be desirable to regularize the estimate $\hat{\beta}$ either by expanding it in terms of a finite number of basis functions, or by attaching an appropriate roughness penalty. These matters are taken up in Chapters 13 and 14, where this model is used to estimate the nonconstant weight functions defining a linear differential operator.

11.7.3 Spline smoothing and other evaluation models

In Section 10.5.2 we discussed some connections between the spline smoothing technique described in Chapter 3 and functional linear modelling. In fact, the spline smoothing model can be put directly into a functional linear modelling context, by somewhat extending the models dealt with in Chapter 10.

Let the dependent variable be a vector $y = (y_1, \ldots, y_N)'$ of N real numbers, and let the model be

$$y_i = \beta(t_i) + \epsilon_i. \tag{11.11}$$

Define the linear mapping

$$z_i(\beta) = \beta(t_i)$$

to transform any function β to its value at t_i. In general, denote by δ_t the *evaluation functional* that maps a function β to its value at t, or

$$\delta_t(\beta) = \beta(t).$$

If we use Dirac delta function notation then we can write $\delta_t(\beta)$ in inner product notation as $\langle \delta_t, \beta \rangle$. One way of considering this notation is to think of the functional δ_t as a delta-function centred at t. The other is simply to regard the inner product notation as a way of writing the effect of the functional δ_t on the function β.

Either way, $z_i = \delta_{t_i}$, so that the model (11.11) can be written as

$$y_i = \langle z_i, \beta \rangle + \epsilon_i. \tag{11.12}$$

Compare this with the model given in equation (10.3); with the exception of the constant term α the model is identical. By allowing more general functionals than inner products with 'nice' functions x_i, we have managed to put spline smoothing into a much more general functional linear modelling framework.

Evaluation models are by no means restricted to simply evaluating the function at sampling values for the argument. We may also consider

$$y_i = D^m \beta(t_i) + \epsilon_i \tag{11.13}$$

where the model is the value of the derivative of order m. Even more generally, we might use

$$\begin{aligned} y_i &= w_0(t_i)\beta(t_i) + w_1(t_i)D\beta(t_i) + \ldots + D^m\beta(t_i) + \epsilon_i \\ &= L\beta(t_i) + \epsilon_i. \end{aligned}$$

Here L is a linear differential operator of order m defined by

$$L = w_0 I + w_1 D + \ldots + D^m = \sum_{j=0}^{m} w_j D^j, \tag{11.14}$$

where the w_j are continuous weight functions and $I\beta = D^0\beta = \beta$. If we then define linear operators z_i to satisfy $\langle z_i, \beta \rangle = L\beta(t_i)$, we have the model (11.12) again.

11.7.4 Weighted and partial integral models

A rather general functional linear model is given by

$$\hat{y}_i(t) = \int \omega_i(s,t) z_i(s) \beta(s,t)\, ds = \langle z_i, \beta(\cdot, t) \rangle_{\omega_i(\cdot,t)} \tag{11.15}$$

where ω_i is a known weight function, and where the notation $\langle \cdot, \cdot \rangle_{\omega(\cdot,t)}$ means taking the inner product using the weight function $\omega(\cdot, t)$.

This setup includes the special case of the partial integral model

$$\hat{y}_i(t) = \int_0^t z_i(s) \beta(s,t)\, ds \tag{11.16}$$

for which

$$\omega(s,t) = \begin{cases} 1, & s \le t \\ 0, & s > t. \end{cases} \tag{11.17}$$

The partial integral model, for example, might be reasonable when we know that the influence of independent variable function z_i extends only backwards in time, as would be the case for nonperiodic processes.

In general, when the weight functions ω vary over argument t, the linear model must be solved pointwise, as is the case for the pointwise multivariate model given above. Again, regularization and basis expansion techniques can be brought into play to ensure that $\hat{\beta}$ is sufficiently smooth and that the estimation problem is manageable.

12

Canonical correlation and discriminant analysis

12.1 Introduction

12.1.1 The basic problem

Suppose we have observed pairs of functions (X_i, Y_i), $i = 1, \ldots, N$, such as the hip and knee angles for the gait cycles of a number of children, as discussed in Section 6.5. We saw there how we can use principal components analysis to examine the variability in the two sets of curves taken together. In this chapter, we pursue a somewhat different emphasis by considering *canonical correlation analysis* (CCA), which seeks to investigate which modes of variability in the two sets of curves are most associated with one another. Thus in the gait analysis example, we might ask how variability in the knee angle cycle is related to that in the hip angle.

First we review classical multivariate CCA; a fuller discussion can be found in most multivariate analysis textbooks, such as Anderson (1984). Then we go on to develop an approach to functional CCA, largely based on the paper of Leurgans, Moyeed and Silverman (1993). The application of the method to two sets of data is presented in Sections 12.3 and 12.4. There we see that some regularization is essential to obtain meaningful results, for reasons discussed briefly in Section 12.5. In Section 12.6, various algorithmic approaches and connections with other FDA topics are explored.

Finally, in Section 12.7, we present some extensions of the ideas of functional CCA to deal with problems of optimal scoring and discriminant analysis. This is based on work of Hastie, Buja and Tibshirani (1995).

12.1.2 Principles of classical canonical correlation analysis

Suppose we have n pairs of observed vectors (x_i, y_i), each x_i being a p-vector and each y_i being a q-vector. Write $\mathbf{V}_{11}$ for the sample variance matrix of the x_i, and $\mathbf{V}_{22}$ for that of the y_i. Let $\mathbf{V}_{12}$ be the $p \times q$ covariance matrix between the x_i and the y_i, so that

$$\mathbf{V}_{12} = \frac{1}{n-1} \sum_{i=1}^{n} (x_i - \bar{x})(y_i - \bar{y})' \tag{12.1}$$

and let $\mathbf{V}_{21} = \mathbf{V}_{12}'$.

The object of canonical correlation analysis is to reduce the dimensionality of the data by finding the vectors a and b (p- and q-vectors respectively) for which the linear combinations $a'x_i$ and $b'y_i$ are as highly correlated as possible. The *canonical variates* $a'x_i$ and $b'y_i$ are the linear compounds of the original observations whose variability is most closely related in terms of correlation. The vectors a and b are called the *canonical variate weight vectors.*

Note that multiplying a and/or b by nonzero constants of the same sign does not alter the correlation. If the constants are opposite in sign, the correlation itself is reversed in sign but has the same magnitude. By convention, we choose a and b, implying that $\{a'x_i\}$ and $\{b'y_i\}$ both have sample variance equal to 1, and the correlation ρ between the $a'x_i$ and $b'y_i$ is positive.

12.1.3 Expressing the analysis as an eigenvalue problem

How can we find a and b? The sample variances of $\{a'x_i\}$ and $\{b'y_i\}$ are $a'\mathbf{V}_{11}a$ and $b'\mathbf{V}_{22}b$ respectively, and the sample covariance of the pairs $(a'x_i, b'y_i)$ is $a'\mathbf{V}_{12}b$. Thus we find the canonical variates by solving the optimization problem

$$\max a'\mathbf{V}_{12}b \text{ subject to } a'\mathbf{V}_{11}a = b'\mathbf{V}_{22}b = 1. \tag{12.2}$$

In Section 12.1.5 we show that this optimization problem can be solved by computing the leading eigenvalue ρ and corresponding eigenvector $\begin{bmatrix} a \\ b \end{bmatrix}$ of the generalized eigenvalue problem

$$\begin{bmatrix} \mathbf{0} & \mathbf{V}_{12} \\ \mathbf{V}_{21} & \mathbf{0} \end{bmatrix} \begin{bmatrix} a \\ b \end{bmatrix} = \rho \begin{bmatrix} \mathbf{V}_{11} & \mathbf{0} \\ \mathbf{0} & \mathbf{V}_{22} \end{bmatrix} \begin{bmatrix} a \\ b \end{bmatrix}. \qquad (12.3)$$

The resulting value of ρ is the correlation between the variates $a'x_i$ and $b'y_i$. Note that the eigenvalues of (12.3) occur in pairs, because if $\begin{bmatrix} a \\ b \end{bmatrix}$ is an eigenvector with eigenvalue ρ, then $\begin{bmatrix} a \\ -b \end{bmatrix}$ is an eigenvector with eigenvalue $-\rho$.

12.1.4 *Subsidiary variates and their properties*

We can go on to solve equation (12.3) completely and find the $r = \text{rank}(\mathbf{V}_{12}) \leq \min(p, q)$ strictly positive eigenvalues $\rho_1 \geq \ldots \geq \rho_r$ with corresponding eigenvectors $(a_1, b_1), \ldots, (a_r, b_r)$. The ρ_j are called the *canonical correlations* of the model. The pairs (a_j, b_j) are the corresponding canonical variate weight vectors; the quantities $(a_j'x_i, b_j'y_i)$ are the pairs of canonical variates, each having sample correlation ρ_j. The canonical variates with the largest correlation ρ_1, as considered in Section 12.1.2, are called the *leading* canonical variates if any confusion is possible.

Arguments from linear algebra found in most multivariate textbooks show that if $j \neq k$ then both of the jth variates are uncorrelated with both of the kth variates. This means that the canonical variates have the following properties for $j \neq k$, where in each case the correlations are the sample correlations as i takes the values $1, \ldots, n$:

(a) $\text{corr}(a_j'x_i, a_k'x_i) = a_j'\mathbf{V}_{11}a_k = 0$

(b) $\text{corr}(b_j'y_i, b_k'y_i) = b_j'\mathbf{V}_{22}b_k = 0$

(c) $\text{corr}(a_j'x_i, b_k'y_i) = a_j'\mathbf{V}_{12}b_k = 0$

In fact, the canonical variates can be characterized successively by maximizing $\text{corr}(a_j'x_i, b_j'y_i)$ subject to the constraints that $a_j'\mathbf{V}_{11}a_k = 0$ and $b_j'\mathbf{V}_{22}b_k = 0$ for $k < j$.

12.1.5 *CCA and the generalized eigenproblem*

In this section, we give a brief justification of the assertion that the solutions of the generalized eigenequation (12.3) yields the solution of the constrained optimization problem (12.4). Readers who are prepared to take this on trust should skip straight to Section 12.2, where we begin our discussion of the functional case.

Consider the problem obtained by weakening the constraint in (12.2) to one on the sum of variances:

$$\max a'\mathbf{V}_{12}b \text{ subject to } a'\mathbf{V}_{11}a + b'\mathbf{V}_{22}b = 2. \tag{12.4}$$

This may be written in block matrix form as

$$\max_{a,b}\left\{[a'\ b']\begin{bmatrix}\mathbf{0} & \mathbf{V}_{12}\\ \mathbf{V}_{21} & \mathbf{0}\end{bmatrix}\begin{bmatrix}a\\ b\end{bmatrix}\right\} \text{ subject to}$$
$$[a'\ b']\begin{bmatrix}\mathbf{V}_{11} & \mathbf{0}\\ \mathbf{0} & \mathbf{V}_{22}\end{bmatrix}\begin{bmatrix}a\\ b\end{bmatrix} = 2. \tag{12.5}$$

By standard optimization arguments, the optimum is obtained by solving the generalized eigenvalue problem (12.3), choosing ρ to be the largest eigenvalue, and then scaling $\begin{bmatrix}a\\ b\end{bmatrix}$ to satisfy the constraint (12.5).

Now suppose that $\begin{bmatrix}a\\ b\end{bmatrix}$ satisfies (12.3). Then $\mathbf{V}_{12}b = \rho\mathbf{V}_{11}a, \mathbf{V}_{21}a = \rho\mathbf{V}_{22}b$ and

$$\rho a'\mathbf{V}_{11}a = a\mathbf{V}_{12}b = b'\mathbf{V}_{21}a = \rho b'\mathbf{V}_{22}b. \tag{12.6}$$

From (12.6) it follows that $\begin{bmatrix}a\\ b\end{bmatrix}$ satisfies the stronger constraint $a'\mathbf{V}_{11}a = b'\mathbf{V}_{22}b = 1$, and hence $\begin{bmatrix}a\\ b\end{bmatrix}$ solves the original optimization problem (12.2) as well as the weaker problem (12.4).

Two other useful properties follow from (12.6). Firstly, the variates $a'x_i$ and $b'y_i$ have correlation ρ, as stated in Section 12.1.3. Secondly, $a'\mathbf{V}_{11}a = b'\mathbf{V}_{22}b$, so whatever scaling is necessary to make the $a'x_i$ have variance one, the same scaling achieves this effect for the $b'y_i$.

12.2 Functional canonical correlation analysis

12.2.1 Notation and assumptions

We now return to the functional case, which is our main concern. As usual, assume that the N observed pairs of data curves (X_i, Y_i) are available for argument t in some finite interval $\mathcal{T}$, and that all integrals are taken over $\mathcal{T}$. Assume that the population mean curves have been estimated and subtracted from the observed data curves, so that sample variance and covariance functions are given as

$$\begin{aligned} v_{11}(s,t) &= N^{-1}\sum X_i(s)X_i(t), \\ v_{22}(s,t) &= N^{-1}\sum Y_i(s)Y_i(t), \text{ and} \\ v_{12}(s,t) &= N^{-1}\sum X_i(s)Y_j(t). \end{aligned}$$

As in Chapter 6 we define corresponding operators V_{11}, V_{22} and V_{12}, by writing $V_{11}f$ for the function

$$V_{11}f(s) = \int_{\mathcal{T}} v_{11}(s,t)f(t)\,dt,$$

and correspondingly for V_{12}, V_{22}. Given functions ξ and η, we define $\texttt{ccorsq}(\xi,\eta)$ to be the sample squared correlation of $\langle \xi, X_i\rangle$ and $\langle \eta, Y_i\rangle$, and therefore

$$\texttt{ccorsq}(\xi,\eta) = \frac{\langle \xi, V_{12}\eta\rangle^2}{\langle \xi, V_{11}\xi\rangle\langle \eta, V_{22}\eta\rangle}.$$

The use of a roughness penalty is central to our methodology. As usual we quantify roughness by integrated squared curvature, imposing periodic boundary conditions if appropriate. For functions f and g satisfying appropriate boundary conditions, we use the property

$$\langle D^2 f, D^2 g\rangle = \langle f, D^4 g\rangle,$$

derived in Section 7.2, and the consequential result that the integrated squared curvature $\|D^2 f\|^2$ can be written $\langle f, D^4 f\rangle$.

12.2.2 *Estimating the leading canonical variates*

For the moment concentrate on the leading canonical variates. We might imagine that the obvious way to proceed is simply to find functions ξ and η that maximize $\texttt{ccorsq}(\xi,\eta)$. This would be equivalent to maximizing $\langle \xi, V_{12}\eta\rangle$ subject to the constraints

$$\langle \xi, V_{11}\xi\rangle = \langle \eta, V_{22}\eta\rangle = 1. \tag{12.7}$$

Both by considering an example and by referring to theoretical results, we shall see later that such an approach breaks down and that maximizing $\texttt{ccorsq}(\xi,\eta)$ does not give any meaningful information about the data or the model. Instead, we must introduce some appropriate smoothing.

A straightforward way of introducing smoothing is to modify the constraints (12.7) by adding roughness penalty terms to give

$$\langle \xi, V_{11}\xi\rangle + \lambda_1\|D^2\xi\|^2 = \langle \eta, V_{22}\eta\rangle + \lambda_2\|D^2\eta\|^2 = 1$$

or, equivalently,

$$\langle \xi, (V_{11} + \lambda_1 D^4)\xi \rangle = \langle \eta, (V_{22} + \lambda_2 D^4)\eta \rangle = 1, \tag{12.8}$$

where λ_1 and λ_2 are positive smoothing parameters.

The effect of introducing the roughness penalty terms into the constraints is that, in evaluating particular candidates to be canonical variates, we consider not only their variances, but also their roughness, and compare a weighted sum of these two quantities with the covariance term. The problem of maximizing the covariance $\langle \xi, V_{12}\eta \rangle$ subject to the constraints (12.8) is equivalent to maximizing the penalized squared sample correlation defined by

$$\texttt{ccorsq}_{(\lambda_1,\lambda_2)}(\xi,\eta) = \frac{\langle \xi, V_{12}\eta \rangle^2}{(\langle \xi, V_{11}\xi \rangle + \lambda_1 \|D^2\xi\|^2)(\langle \eta, V_{22}\eta \rangle + \lambda_2 \|D^2\eta\|^2)}. \tag{12.9}$$

We refer to this procedure as *smoothed canonical correlation analysis.*

The larger the values of λ_1 and λ_2, the more emphasis is placed on the roughness penalty and the smaller will be the true correlation of the variates found by smoothed CCA. A good choice of the smoothing parameters is essential to give a pair of canonical variates with fairly smooth weight functions and a correlation that is not unreasonably low.

By an argument analogous to that set out for the multivariate case in Section 12.1.3, the functions (ξ, η) that maximize equation (12.9) are the eigenfunctions corresponding to the largest positive eigenvalue ρ of the system of operator equations

$$\begin{pmatrix} 0 & V_{12} \\ V_{21} & 0 \end{pmatrix} \begin{pmatrix} \xi \\ \eta \end{pmatrix} = \rho \begin{pmatrix} V_{11} + \lambda_1 D^4 & 0 \\ 0 & V_{22} + \lambda_2 D^4 \end{pmatrix} \begin{pmatrix} \xi \\ \eta \end{pmatrix}. \tag{12.10}$$

Some details of practical ways for dealing with this equation are given in Section 12.6.

Generally, we have found it sufficient to consider the special case where $\lambda_1 = \lambda_2 = \lambda$, and for the practical discussion of functional CCA in the next two sections we confine attention to this case, using the single subscript λ where necessary. The extension to the more general case is straightforward.

Our method of introducing smoothing or regularization is similar to the technique of ridge regression, which is often used in image processing and ill-posed problems to improve the conditioning of covariance matrices corresponding to V_{11} and V_{22}. The technique of ridge regression was applied to CCA by Vinod (1976).

12.2.3 *Cross-validation and subsidiary variates*

The smoothing parameter λ in the procedure set out can be chosen subjectively. If we require an automatic procedure, a reasonable form of cross-validation is as follows:

Let $\mathtt{ccorsq}_\lambda^{-i}(\xi, \eta)$ be the sample penalized squared correlation calculated as in (12.9) but with the observation (X_i, Y_i) omitted. Let $(\xi_\lambda^{(-i)}, \eta_\lambda^{(-i)})$ be the functions that maximize $\mathtt{ccorsq}_\lambda^{-i}(\xi, \eta)$. The cross-validation score for λ is defined to be the squared correlation of the N pairs of numbers

$$(\langle \xi_\lambda^{(-i)}, X_i \rangle, \langle \eta_\lambda^{(-i)}, Y_i \rangle)$$

for $i = 1, \ldots, n$. We then choose the value of λ that maximizes this correlation.

Throughout this section, we have concentrated on estimating the leading canonical variates. Of course, the subsequent positive eigenvalues and corresponding eigenfunctions of system (12.10) give the subsidiary canonical correlations and their canonical variates. These are orthogonal with respect to the *penalized* sample covariance operators $V_{11} + \lambda D^4$ and $V_{22} + \lambda D^4$, rather than the raw sample covariance operators as in the classical case. If we were particularly interested in the ideal smoothing parameter for a subsidiary canonical correlation, we could formulate a relevant cross-validation score. However, our practical experience has shown us that, although cross-validation works well for the leading canonical variate, its behaviour is much more disappointing for subsequent canonical variates.

12.3 Application to the gait data

We can now apply the canonical correlation analysis approach to the gait data as discussed in Chapters 1 and 6, considering both the hip and the knee angles.

Figure 12.1 shows functions ξ and η that maximize the *unsmoothed* sample correlation $\mathtt{ccorsq}$. The sample correlation achieved by these functions is 1. The functions displayed in Figure 12.1 do not give any meaningful information about the data and clearly demonstrate the need for a technique involving smoothing. In Section 12.5, we explain why this behaviour is not specific to this particular data set but is an intrinsic property of CCA applied in the functional context.

In Figure 12.2, we see the leading smoothed canonical variates ξ_λ and η_λ with the smoothing parameter λ chosen by the cross-validation method described in Section 12.2.3. Since the main interest

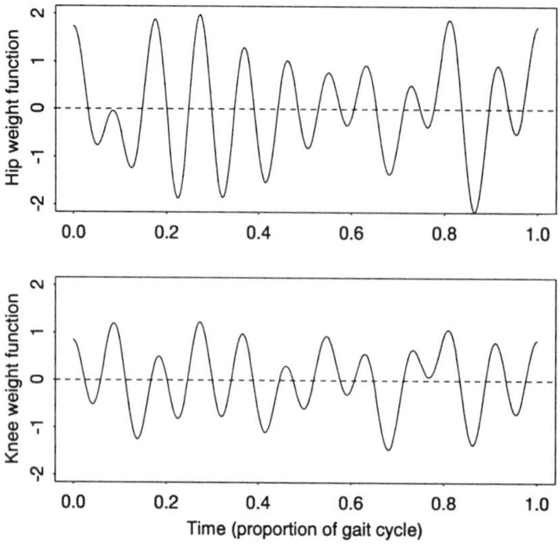

FIGURE 12.1. Unsmoothed canonical variate weight functions for the gait data that attain perfect correlation. Top panel: weight function for hip observations; bottom panel: weight function for knee observations.

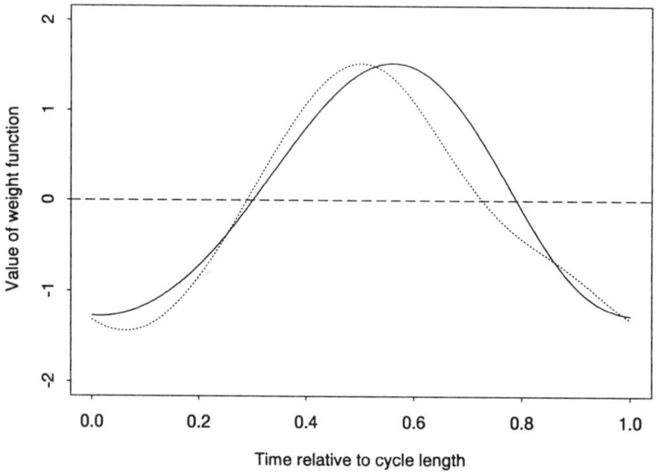

FIGURE 12.2. First pair of smoothed canonical variate weight functions for the gait data. Solid curve: weight function for hip observations; dotted curve: weight function for knee observations.

TABLE 12.1. Smoothed and unsmoothed sample correlations for the first three pairs of smoothed canonical variates for the gait data.

Canonical variates	Sample squared correlations $\mathrm{ccorsq}_\lambda(\xi_\lambda, \eta_\lambda)$	$\mathrm{ccorsq}(\xi_\lambda, \eta_\lambda)$
First	0.755	0.810
Second	0.618	0.717
Third	0.141	0.198

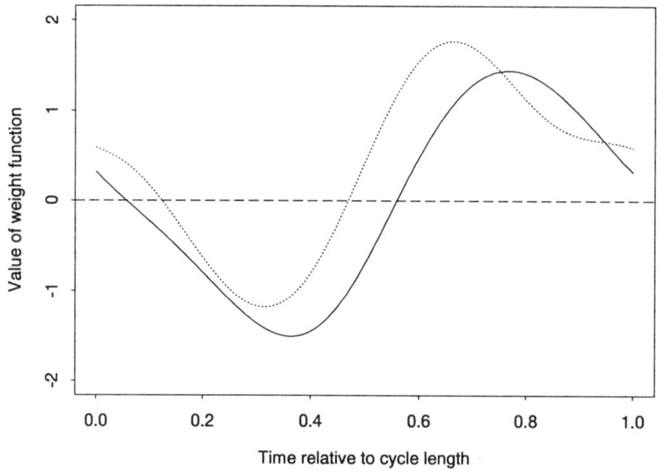

FIGURE 12.3. Second pair of smoothed canonical variate weight functions for the gait data. Solid curve: weight function for hip observations; dashed curve: weight function for knee observations.

is in comparing the curves, they have been normalized to set the integral of their squares equal to 1. Table 12.1 shows the values of the smoothed and unsmoothed squared correlations, and also includes corresponding values for the second and third pairs of smoothed canonical variates, estimated with the same λ. The broad interpretation of these variates is that there is correlation between the two measurements at any particular time. But it is interesting that the extreme in the hip curve in the middle of the cycle occurs a little later than that in the knee curve, whereas the order of the extremes near the beginning of the cycle is reversed. This suggests that, in the middle of the cycle, high variability from the norm in the hip follows that in the knee; near the ends of the cycle, the effects occur in the opposite order. This may indicate a physical propagation of errors caused by the relevant strike of the heel at the beginning and in the middle of the cycle.

Table 12.1 shows that the second pair of canonical variates is almost as important as the first, and these are displayed in Figure 12.3, with the same normalization as in Figure 12.2. The points at which the second canonical variates cross the axis indicate conclusions similar to those outlined with respect to the leading variates. In the middle of the cycle the hip curve crosses zero considerably later than the knee curve, whereas near the beginning of the cycle the hip curve crosses first. Put another way, we could roughly transform both the first and the second canonical variates to be identical for the hip and the knee by speeding up the hip cycle relative to the knee cycle in the first half of the cycle, and slowing it down in the second. The degree of smoothing chosen by cross-validation appears to be quite heavy, and to test the sensitivity of these conclusions, Leurgans, Moyeed and Silverman (1993) examined the first two pairs of canonical variates estimated with a value of λ reduced by a factor of 10. Though there was a little more variability in the canonical variate curves, the broad features remained the same.

Table 12.1 shows that the values of the estimated correlations of the third pair of canonical variates are small, and for our purposes can be ignored. Scatterplots of the canonical variate scores $(\langle \xi, X_i \rangle, \langle \eta, Y_i \rangle)$ show that no particular curves have outlying scores for either of the first two canonical variates.

In Section 6.5, we saw that the first principal component of variation in the hip curves alone corresponded to an overall vertical shift in the curves. If this shift were in any way correlated with a variation in the knee curves, the hip canonical variate curves would be more like constants than sine waves. Since this is not the case, we can see that this vertical shift is a property of the hip curves alone, independent of any variation in the knee angles.

12.4 Application to the study of lupus nephritis

Buckheit, Olshen, Blouch and Myers (1997) applied functional CCA to renal physiology, in the study of diffuse proliferative lupus nephritis, and we present their results here as an illustration. The original paper should be consulted for further details; we are extremely grateful to Richard Olshen for his generosity in sharing and discussing this work with us prior to its publication.

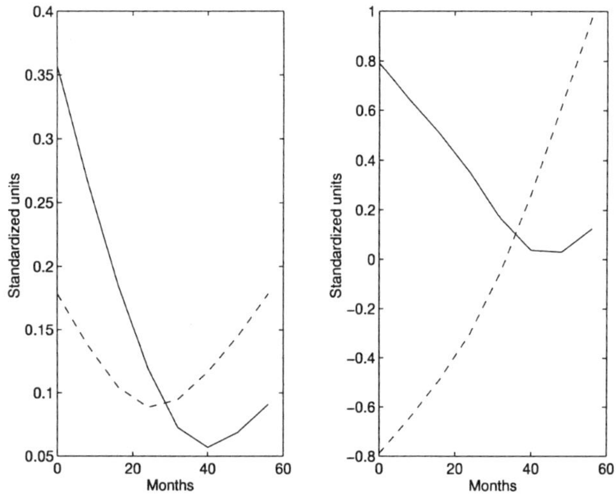

FIGURE 12.4. Smoothed canonical variate weight functions for the lupus data, from Buckheit *et al.* (1997). Left panel: results of CCA applied to GFR and KUC with solid curve corresponding to GFR and dashed curve to KUC. Left panel: results of CCA applied to GFR and GOP with solid curve corresponding to GFR and dashed curve to GOP.

They had available various measurements on a number of patients over a 60 month period. These include the glomerular filtration rate (GFR), the glomerular oncotic pressure (GOP) and the two-kidney ultrafiltration coefficient (KUC). They focused on nine patients labelled *progressors*, those whose kidney function, as measured by GFR was clearly declining over the period of study. The GFR measure is currently favoured by clinicians as an overall indicator of progressive glomerular disease, a particular form of kidney degeneration, and therefore the progressors are the group suffering long-term kidney damage, likely to require eventual dialysis or transplantation. It is important to understand the kidney filtration dynamics in this disease, and this is facilitated by investigating the covariation between measured variables.

Within the progressor group, GFR and KUC tend to decrease considerably over the 60 month period, whereas the GOP measure increases somewhat. This contrasts with well-functioning kidneys, where an increase in GOP would be counteracted by an increase in KUC resulting in steady GFR. Functional smoothed CCA was applied to explore variability and interaction effects in the progressor group. The correlations between GFR and each of KUC and GOP were investigated.

Figure 12.4 shows the leading pairs of canonical variate weight functions. It is interesting that the linear functional of GFR most highly correlated with the other two variables is virtually the same in both cases.

To interpret the figure, remember that all patients concerned show an overall declining value of GFR. The U-shaped solid curves in the figure therefore correspond to a canonical variate where a positive value indicates a GFR record that starts at a value higher than average, but then declines more rapidly than average in the first 40 months, finally switching to a relatively less rapid decline in the last 20 months.

The left hand panel shows that this variate is correlated with a similar effect for KUC, but the switch in rate happens earlier. This indicates not only that strong decline of GFR is associated with strong decline of KUC but also suggests that the pattern of GFR in some sense follows that of KUC, raising the hope that KUC could be used to predict future GFR behaviour. On the other hand, the right hand panel shows that this aspect of GFR behaviour is correlated with an increase of GOP stronger than average over the entire time period. Thus, patients with rapidly increasing GOP are likely to be those whose GFR declines rapidly at first, though there may be some reduction in the rate of decline after about 36 months.

In broad terms, the CCA gives insights broadly consistent with those for the average behaviour of the sample as a whole. It is interesting that the relationships between the variables are borne out on an individual level, not merely on an average level. Furthermore the detailed conclusions yielded by the CCA give important avenues for future thought and investigation concerning the way in which the variables interrelate. Of course, given the small sample size, any conclusions must be relatively tentative unless supported by other evidence.

12.5 Why is regularization necessary?

Apart from its importance as a practical method, canonical correlation analysis of functional data has an interesting philosophical aspect. In the principal components analysis context we have already seen that appropriately applied smoothing may improve the estimation accuracy. However, in most circumstances, we obtain reasonable estimates of the population principal components even if no smoothing is applied. By contrast, as we saw in the gait example, in the context of functional CCA some regularization is absolutely essential to obtain meaningful results. This is the same conclusion that we drew for the functional regression context discussed in Chapter 10. But in the canonical correlation case, the impact of smoothing is even more dramatic.

To understand the need for regularization, compare functional CCA with standard multivariate CCA. A standard condition of classical CCA is that $n > p+q+1$ which ensures (with probability 1, under reasonable conditions) that the sample covariance matrix $\mathbf{V}_{12}$ of the n vectors (x_i, y_i) is nonsingular (see Eaton and Perlman,1973). In the functional case, p and q are essentially infinite, and so this condition cannot be fulfilled.

Furthermore, consider a sample $X_1, \ldots, X_N$ of functional data, and assume for the moment that the N curves are linearly independent. Now suppose that $z_1, \ldots, z_N$ is any real vector. By the results of Chapter 10, there is a functional interpolant ξ such that, for some constant α_X, $z_i = \alpha_X + \langle \xi, X_i \rangle$ for all i. Now suppose we have a second sample of curves Y_i, which may be correlated with the X_i in some way, and again are linearly independent. We can also find a functional interpolant η such that, for some constant α_Y, $z_i = \alpha_Y + \langle \eta, Y_i \rangle$ for all i. This means that the given values z_i can be predicted perfectly either from the X_i or from the Y_i.

It follows that not only have we found functions ξ and η such that $\mathtt{ccorsq}(\xi, \eta) = 1$, because the variates $\langle \xi, X_i \rangle$ and $\langle \eta, Y_i \rangle$ are perfectly correlated, but that we can prescribe the values z_i taken by the canonical variates to be whatever we please, up to a constant. In particular, we could start with *any* function ξ, construct $z_i = \langle \xi, X_i \rangle$, and then find a function η such that $\mathtt{ccorsq}(\xi, \eta) = 1$. In this sense, every possible function can arise as a canonical variate weight function with perfect correlation!

Leurgans, Moyeed and Silverman (1993) discuss this result in greater detail. They demonstrate that the assumption of linear independence among the curves is a very mild one, and, by proving an appropriate consistency result, they show that regularization indeed makes meaningful estimates possible.

12.6 Algorithmic considerations

12.6.1 Discretization and basis approaches

Just as in the case of functional regression discussed in Chapter 10, there are several ways of carrying out our method of smoothed functional CCA numerically. For completeness, we present the methodology for the general case of different parameters λ_1 and λ_2. A direct approach is to set up a discrete version of the eigenequation (12.10). Discretize the functions ξ and η and the covariance operators $v_{jk}(s,t)$ using a fine grid, and replace the operator D^4 by a finite difference approximation. System (12.10) then becomes a large linear system whose leading eigenvalue and eigenvector are found by standard numerical methods.

We can also use a basis for the functions X_i and Y_i, and for the weight functions ξ and η. Suppose that $\phi_1, \phi_2, \ldots, \phi_M$ is a suitable basis as in Sections 10.4.1 and 10.6.2. As usual, define $\mathbf{K}$ to be the matrix with entries $\langle D^2\phi_j, D^2\phi_k\rangle$ and $\mathbf{J}$ the matrix with entries $\langle \phi_j, \phi_k\rangle$. If we use a Fourier or other orthonormal basis, then $\mathbf{J}$ is the identity matrix.

Define $\mathbf{C}$ and $\mathbf{D}$ to be the matrices of coefficients of the basis expansions of the X_i and Y_i respectively, meaning that

$$X_i = \sum_{\nu=1}^{M} c_{i\nu}\phi_\nu$$

and

$$Y_i = \sum_{\nu=1}^{M} d_{i\nu}\phi_\nu$$

up to the degree of approximation involved in any choice of the number M of basis functions considered. Write a and b for the vectors of coefficients of the basis expansions of the functions ξ and η.

Define $M \times M$ covariance matrices $\tilde{\mathbf{V}}_{11}$, $\tilde{\mathbf{V}}_{12}$ and $\tilde{\mathbf{V}}_{22}$ to be the matrices with (ν, ρ) entries

$$N^{-1}\sum_i c_{i\nu}c_{i\rho},\ N^{-1}\sum_i c_{i\nu}d_{i\rho},\ \text{and } N^{-1}\sum_i d_{i\nu}d_{i\rho},$$

respectively, the sample variance and covariance matrices corresponding to the basis expansions of the data. It is straightforward to show that, in the basis expansion domain, we carry out the smoothed CCA of the given data by solving the generalized eigenvalue problem

$$\begin{bmatrix} \mathbf{0} & \mathbf{J}\tilde{\mathbf{V}}_{12}\mathbf{J} \\ \mathbf{J}\tilde{\mathbf{V}}_{21}\mathbf{J} & \mathbf{0} \end{bmatrix} \begin{bmatrix} a \\ b \end{bmatrix} = \rho \begin{bmatrix} \mathbf{J}\tilde{\mathbf{V}}_{11}\mathbf{J} + \lambda_1\mathbf{K} & \mathbf{0} \\ \mathbf{0} & \mathbf{J}\tilde{\mathbf{V}}_{22}\mathbf{J} + \lambda_2\mathbf{K} \end{bmatrix} \begin{bmatrix} a \\ b \end{bmatrix}.$$

As in Chapter 10, we should choose the number of basis functions M large enough to ensure that the regularization is controlled by the choice of the smoothing parameter(s) λ rather than that of dimensionality M. Values of M of around 20 should give good results without imposing an excessive computational burden.

12.6.2 The roughness of the canonical variates

A third algorithmic possibility is related to the idea of quantifying of the roughness of a variate, as discussed in Section 10.5.1. Just as in the case of functional regression, this idea is of both conceptual and algorithmic value, and can be used to elucidate the regularization method we propose for functional canonical correlation analysis.

Suppose $z_i = \langle \xi, X_i \rangle$ is a possible canonical variate. Let $\mathbf{R}_X$ be the matrix $\mathbf{R}$ as derived in Section 10.7.3, implying that $z'\mathbf{R}_X z$ is the roughness of the smoothest functional interpolant to the values z_i with respect to the functional covariates X_i. It may be that this value is equal to $\|D^2\xi\|^2$, or it may be that z_i can be obtained by integrating a smoother function against the X_i. In any case, we can consider $z'\mathbf{R}_X z$ in its own right as a measure of the roughness of z_i as a variate based on the X_i.

Similarly, let $\mathbf{R}_Y$ be a matrix such that the roughness of any variate w relative to the observed covariate functions $\{Y_i\}$ is $w'\mathbf{R}_Y w$. Our smoothed canonical correlation method can then be recast as the determination of variates z and w to maximize the sample covariance of z_i and w_i subject to

$$\mathtt{var}\{z_i\} + \lambda_1 z'\mathbf{R}_X z = \mathtt{var}\{w_i\} + \lambda_2 w'\mathbf{R}_Y w = 1. \tag{12.11}$$

Once we have found in this way a pair of canonical variates, the corresponding weight functions are defined as the smoothest functions ξ and η satisfying $z_i = \langle \xi, X_i \rangle$ and $w_i = \langle \eta, Y_i \rangle$ for all i. Since we are concerned only with the variance and covariance of the $\{z_i\}$ and $\{w_i\}$, we can discard the constant terms in the functional interpolation.

As before, we can maximize the sample covariance of $\{z_i\}$ and $\{w_i\}$ subject to the constraints (12.11) by solving an eigenvalue problem. Some care is necessary to deal with a slight complication caused by the presence of the sample mean in the formula for variance and covariance.

Assuming without loss of generality that the canonical variates have sample mean zero, write the constrained maximization problem as that of finding the maximum of $z'w$ subject to the constraints

$$z'z + \lambda_1 z'\mathbf{R}_X z = w'w + \lambda_2 w'\mathbf{R}_Y w = 1 \tag{12.12}$$

and the additional constraints

$$1'z = 1'w = 0. \tag{12.13}$$

For the moment, neglect the constraint (12.13) and consider the maximization of $z'w$ subject only to the constraints (12.12). This corresponds to the eigenvalue problem

$$\begin{bmatrix} \mathbf{0} & \mathbf{I} \\ \mathbf{I} & \mathbf{0} \end{bmatrix} \begin{bmatrix} z \\ w \end{bmatrix} = \rho \begin{bmatrix} \mathbf{I} + \lambda_1 \mathbf{R}_X & \mathbf{0} \\ \mathbf{0} & \mathbf{I} + \lambda_2 \mathbf{R}_Y \end{bmatrix} \begin{bmatrix} z \\ w \end{bmatrix}. \tag{12.14}$$

By premultiplying (12.14) by $[z' \;\; w']$ and taking the product of the two expressions for $z'w$ thus obtained, any solution of (12.14) satisfies

$$(z'w)^2 = \rho^2 (z'z + \lambda_1 z' \mathbf{R}_X z)(w'w + \lambda_2 w' \mathbf{R}_Y w) \geq \rho^2 (z'z)(w'w)$$

and so it is necessarily the case that $|\rho| \leq 1$. Since the smoothest functional interpolant of the constant vector has roughness zero, $\mathbf{R}_X 1 = \mathbf{R}_Y 1 = 0$, and so the condition $z = w = 1$ yields the leading solution of (12.14), with eigenvalue $\rho = 1$.

The solution of (12.14) with the *second* largest eigenvalue maximizes $z'w$ subject to the constraint (12.12) and the additional constraint

$$1'(\mathbf{I} + \lambda_1 \mathbf{R}_X) z = 1'(\mathbf{I} + \lambda_2 \mathbf{R}_Y) w = 0. \tag{12.15}$$

But since $\mathbf{R}_X 1 = \mathbf{R}_Y 1 = 0$, the constraint (12.15) is precisely equivalent to the constraint (12.13) that we temporarily neglected. It follows that the second and subsequent eigensolutions of (12.14) are the canonical variates we require, and automatically have sample mean zero; the leading solution is a constant and should be ignored.

12.7 Penalized optimal scoring and discriminant analysis

Hastie, Buja and Tibshirani (1995) consider functional forms of the multivariate techniques of optimal scoring and linear discriminant analysis, making use of ideas closely related to the functional canonical correlation analysis approach discussed in this chapter. We present a brief overview of their work; see the original paper for further details.

12.7.1 The optimal scoring problem

Assume that we have N paired observations (X_i, y_i) where each X_i is a function, and each y_i is a category or class taking values in the set $\{1, 2, \ldots, J\}$. For notational convenience, we code each y_i as a J-vector with value 1 in position j if $y_i = j$, and 0 elsewhere.

We aim to obtain a function β and a J-vector θ minimizing the criterion

$$\mathsf{OSERR}(\theta, \beta) = N^{-1} \sum_{i=1}^{N} (\langle \beta, X_i \rangle - \theta' y_i)^2$$

subject to the normalization constraint $N^{-1}\sum_i(\theta' y_i)^2 = 1$. The idea is to turn the categorical variable coded by the y-vectors into a quantitative variable taking the values θ_j. The θ_j are the scores for the various categories, chosen to give the best available prediction of a linear property $\langle \beta, X \rangle$ of the observed functional data.

For any given θ, the problem of finding the functions β is the functional regression problem discussed in Chapter 10. There we saw that, without some regularization, it is usually possible to choose β to give perfect prediction of any specified values $\theta' y_i$. This means that we cannot choose an optimal score vector θ uniquely on the basis of the observed data. To deal with this difficulty, Hastie et al. (1995) introduced the *penalized* optimal scoring criterion

$$\mathsf{OSERR}_\lambda(\theta, \beta) = \mathsf{OSERR}(\theta, \beta) + \lambda \times \mathsf{PEN}(\beta),$$

where λ is a smoothing parameter and $\mathsf{PEN}(\beta)$ a roughness penalty.

12.7.2 The discriminant problem

The discriminant problem is similar to the optimal scoring problem. Again, we have functional observations X_i, each allocated to a category in $\{1, 2, \ldots, J\}$. For any proposed linear discriminant functional $\langle \beta, X_i \rangle$, define θ_j to be the average of the $\langle \beta, X_i \rangle$ for all X_i falling in category j. For each fixed β, this value of θ minimizes the quantity $\mathsf{OSERR}(\theta, \beta)$, which can then be re-interpreted as the *within-class variance* of the $\langle \beta, X_i \rangle$. The *between-class variance* is simply the variance of the discriminant class means $\theta' y_i$, defining the J-vectors y_i by the same coding as above. Discriminant analysis aims to maximize the between-class variance subject to a constraint on the within-class variance.

The roles of objective function and constraint are exchanged in passing from optimal scoring to discriminant analysis, and minimization is replaced by maximization. Also, primary attention shifts from the score vector θ in optimal scoring to the discriminant functional defined by the function β in discriminant analysis. Hastie et al. make the correspondence complete by proposing *penalized discriminant analysis* where we maximize the raw between-class variance subject to a penalized constraint on the within-class variance

$$\mathsf{OSERR}(\theta, \beta) + \lambda \times \mathsf{PEN}(\beta) = 1.$$

12.7.3 The relationship with CCA

Simple modifications of arguments from multivariate analysis show that the penalized optimal scoring and the penalized discriminant analysis problems are both equivalent to the mixed functional-multivariate canonical correlation analysis problem of maximizing the covariance of $\langle \xi, X_i \rangle$ and $\eta' y_i$ subject to the constraints

$$\text{var}(\langle \xi, X_i \rangle) + \lambda \times \text{PEN}(\xi) = \text{var}(\eta' y_i) = 1. \tag{12.16}$$

In the notation we have used for CCA, the weight corresponding to the functional part X_i of the data is itself a function ξ, whereas the vector part y_i is mapped to its canonical variate by a weight *vector* η. Only the functional part ξ is penalized for roughness in the constraints (12.16).

This formulation gives rise to an eigenequation of the form (12.10) but with no penalty attached to V_{22}. The numerical approaches we have set out for CCA carry over to this case, with appropriate modifications because only the X_i are functions.

To obtain the solutions (β, θ) of the discriminant and optimal scoring problems, it is only necessary to rescale the estimated function ξ and vector η appropriately. The subsidiary eigensolutions, as mentioned in Section 12.2.3, are also interesting for these problems because they yield estimates of vector-valued scores θ_j and discriminants $\langle \beta, X_i \rangle$.

12.7.4 *Applications*

Hastie et al. present two fascinating applications of these techniques. For speech recognition, the frequency spectra of digitized recordings of various phonemes are used as data. A roughness penalty of the form $\text{PEN}(\beta) = \int \{D^2\beta(\omega)\}^2 w(\omega) d\omega$ is used, with the weight function $w(\omega)$ chosen to place different emphasis on different frequencies ω.

Their other application is the recognition of digits in handwritten postal addresses and zip codes. In this case, the observations X_i are functions of a bivariate argument t, defined in practice on a 16×16 pixel grid. The roughness penalty used is a discrete version of the Laplacian penalty $\iint [\nabla^2\beta(t)]^2\, dt$.

13

Differential operators in functional data analysis

13.1 Introduction

The derivatives of functional observations have played a strong role from the beginning of this book. For example, for growth curves and handwriting coordinate functions we chose to work with acceleration directly, and for temperature profiles we considered functions $(\pi/6)^2 D\,\mathtt{Temp} + D^3\mathtt{Temp}$. We used $D^2\beta$ to construct a measure of curvature in an estimated regression function β to regularize or smooth the estimate, and applied this same principle in functional principal components analysis, canonical correlation, and other types of linear models. Thus, derivatives can be used both as the object of inquiry and as tools for stabilizing solutions.

It is time to look more systematically at how derivatives might be employed in FDA. Are there other ways of using derivatives, for example? Can we use mixtures of derivatives instead of simple derivatives? Can we extend models so that derivatives can be used on either the covariate or response side? Can our smoothing and regularization techniques be extended in useful ways? Are new methods of analysis making explicit use of derivative information possible?

The next two chapters explore these questions. But since certain results drawn from the theory of differential equations are essential

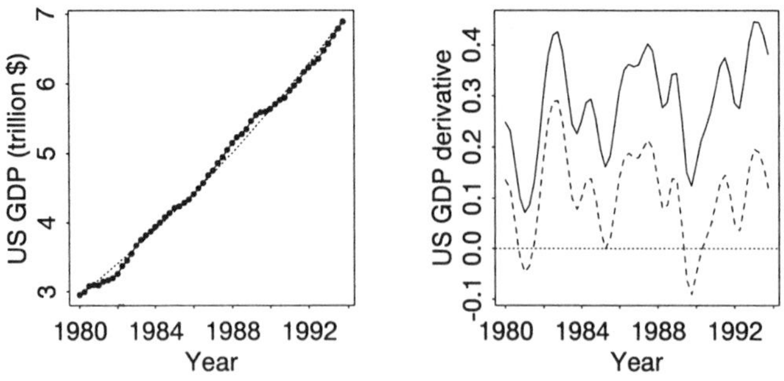

FIGURE 13.1. The left panel shows the gross domestic product of the United States in trillion US$. The solid curve is a polynomial smoothing spline constructed with penalty the integrated squared fourth derivative, and the dotted curve is a purely exponential trend fit by least squares. The solid curve in the right panel is the estimated first derivative of GDP. The dashed curve in this panel is the value of the differential operator $L = -\gamma \, \mathrm{GDP} + D \, \mathrm{GDP}$.

to this enquiry, first of all we present and explain, without proof, some essential preliminary results.

The central idea to be explored is that of a *linear differential operator* L, which is a weighted combination of derivatives up to a specified order m:

$$L = w_0 I + w_1 D + w_2 D^2 + \ldots + D^m \tag{13.1}$$

where the m coefficient functions $w_j, j = 0, \ldots, m-1$ define the operator. For brevity we make free use of the acronym LDO for linear differential operator.

First, however, we introduce some data to illustrate these notions.

13.2 Two sets of data

13.2.1 *The gross domestic product data*

The gross domestic product (GDP) of a country is the financial value of all goods and services produced in that country, whether by the private sector of the economy or by government. Like most economic measures, GDP tends to exhibit a percentage change each year in times of domestic and international stability. Although this change can fluctuate considerably from year to year, over long periods the fluctuations tend to even out for most countries and the long-range trend in GDP tends to be roughly exponential.

We obtained quarterly GDP values for 15 countries in the Organization for Economic Cooperation and Development (OECD) for the years 1980 through 1994 (OECD, 1995). The values for any country are expressed in its own currency, and thus scales are not comparable across countries. Also, there are strong seasonal effects in GDP values reported by some countries, whereas others smooth them out before reporting.

The left panel of Figure 13.1 displays the GDP of the United States. The seasonal trend, if any, is hardly visible, and the solid line indicates a smooth of the data using a penalty on D^4 GDP. It also shows a best fitting exponential trend, $C \exp(\gamma t)$, with rate constant $\gamma = 0.038$. Thus, over this period the U.S.. economy tended to grow at about 4% per year. The right panel displays the first derivative of GDP as a solid line. The economy advanced especially rapidly in 1983, 1987 and 1993, but there were slowdowns in 1981, 1985 and 1990.

13.2.2 *The melanoma data*

Figure 13.2 presents age-adjusted melanoma incidences for 37 years from the Connecticut Tumor Registry (Houghton et al., 1980). Two types of trend are obvious: a steady linear increase and a periodic component. The latter is related to sunspot activity and the accompanying fluctuations in solar radiation. Two smooths of these data are given: the solid line is a polynomial smoothing spline penalizing the squared norm of the fourth derivative $D^4 x$ with penalty parameter chosen by minimizing generalized cross-validation, and the dotted curve is a least squares fit of the function

$$x(t) = c_1 + c_2 t + c_3 \sin(\omega t) + c_4 \cos(\omega t). \tag{13.2}$$

Clearly, this linear/periodic function fits the data well, but is there additional variation to be estimated? The polynomial spline fit differs slightly from this function, but a choice of differential operator L to define the penalty term that would be more natural than $L = D^4$ is

$$Lx = \omega^2 D^2 x + D^4 x$$

since, given the appropriate value for period $2\pi/\omega$, functions in the class (13.2) satisfy $Lx = 0$ and would therefore receive zero penalty.

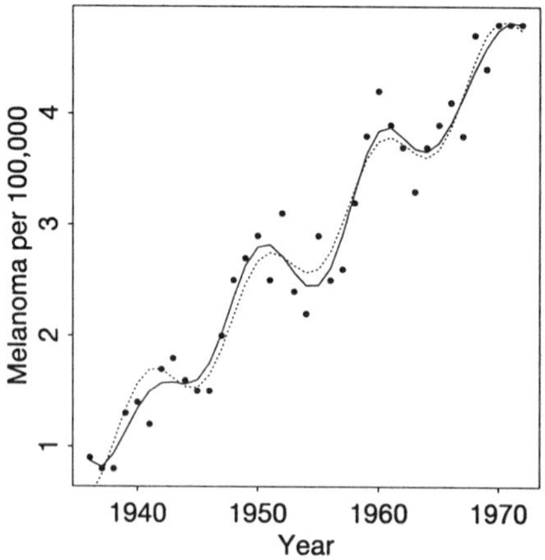

FIGURE 13.2. Age-adjusted incidences of melanoma for the years 1936 to 1972. The solid curve is the polynomial smoothing spline fit to the data penalizing the norm of the fourth derivative, with the smoothing parameter chosen by minimizing the GCV criterion. The dotted line is the fit of $x(t) = c_1 + c_2 t + c_3 \sin(\omega t) + c_3 \cos(\omega t)$ with $\omega = 0.65$ and coefficients estimated by least squares.

13.3 Some roles for linear differential operators

13.3.1 Operators to produce new functional observations

Derivatives of various orders and mixtures of them are of immediate interest in many applications. We have already noted that there is much to be learned about human growth by examining acceleration profiles. There is an analogy with mechanical systems; a version of Newton's third law, $a(t) = F(t)/M$, asserts that the application of some force $F(t)$ at time t on an object with mass M has an immediate impact on acceleration $a(t)$. However, it has only an indirect impact on velocity, $v(t) = v_0 + M^{-1} \int_0^t F(u)du$, and an even less direct impact on what we directly observe, namely position $s(t) = s_0 + v_0 t + M^{-1} \int_0^t \int_0^u F(z)dz$. From the standpoint of mechanics, the world that we experience is two integrals away from reality! The release of adrenal hormones during puberty tends to play the role of the force function F, and so does a muscle contraction with respect to the position of a limb or other part of the body.

Consider another kind of growth, exponential growth or decay exhibited by systems such as populations, radioactive particles and economic indicators with arbitrary value α at the origin,

$$x(t) = \alpha e^{\gamma t}. \tag{13.3}$$

If we take the ratio

$$\frac{Dx}{x} = D \ln x = \gamma,$$

we can isolate the rate parameter γ, and this can also be inspected by plotting Dx against x and computing the slope. Put yet another way, an exponential growth process of this nature satisfies the simple linear differential equation

$$\gamma x = Dx \tag{13.4}$$

or, alternatively, the linear homogeneous differential equation

$$-\gamma x + Dx = 0. \tag{13.5}$$

If we define Lx to be $-\gamma x + Dx$, then we may say even more compactly that $Lx = 0$ implies exponential growth. When studying processes that exhibit exponential growth or decay to some extent, it can therefore be helpful to look at Lx defined in this way; the extent to which the result is a nonzero function with substantial variation is a measure of departure from exponential growth, just as the appearance of a nonzero phase in D^2x for a mechanical system indicates the application of a force.

For the GDP data, one of several techniques for estimating the rate parameter γ is to fit model (13.3) to data by least squares. This produces an estimate of 0.038 for the U.S. data displayed in Figure 13.1, and the fitted exponential trend is shown there as the dotted curve in the left panel. The right panel presents the values of $L\,\text{GDP} = -0.038\,\text{GDP} + D\,\text{GDP}$ as the dashed curve. This function indicates the periods of departure from exponential growth, and makes the hypergrowth epochs in 1983, 1987 and 1993, and recessions in 1981 and 1990 even more apparent. Figure 13.3 shows the comparable curves for the United Kingdom and Japan, and we note that the U.K. had only one boom period with an uncertain recovery after the recession, whereas Japan experienced a deep and late recession.

13.3.2 Operators to regularize or smooth models

Although we have covered this topic elsewhere, we should still point out that we may substitute Lx for D^2x in any of the regularization schemes covered so far. Why? The answer lies in the homogeneous

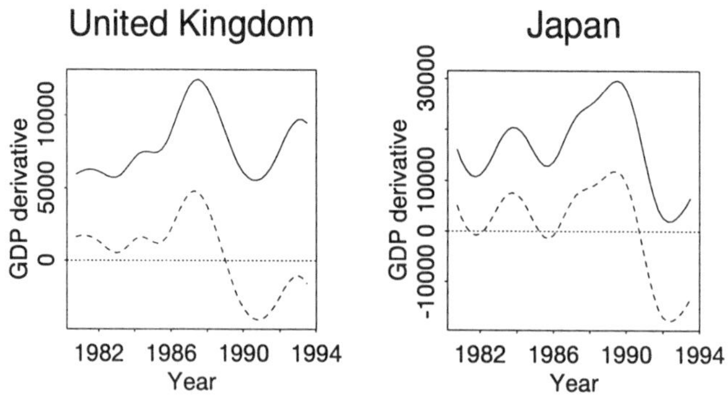

FIGURE 13.3. The solid curves are the derivatives of GDP of the United Kingdom and Japan estimated by smoothing splines penalizing for integrated squared fourth derivative. The dashed curves are the corresponding values of the differential operator $L = -\gamma x + Dx$.

equation $Lx = 0$; functions satisfying this equation are deemed to be hypersmooth in the sense that we choose to ignore any component of variation of this form in calculating roughness or irregularity. In the case of the operator D^2, linear trend is considered to be so smooth that any function may have an arbitrary amount of it, since the penalty term $\lambda \int (D^2 x)^2$ is unaffected. On the other hand, suppose that we are working with a process that has predominantly exponential growth with rate parameter γ. In this case we may choose nonparametric regression with the fitting criterion

$$\text{PENSSE}_\lambda(x) = n^{-1} \sum_{j=1}^{n} [y_j - x(t_j)]^2 + \lambda \int [-\gamma x(t) + Dx(t)]^2 \, dt$$

to leave untouched any component of variation of this form.

More generally, suppose we observe a set of discrete data values generated by the process

$$y_j = x(t_j) + \epsilon_j$$

where, as in previous chapters, x is some unobserved smooth function that we wish to estimate by means of nonparametric regression, and ϵ_j is a disturbance or error assumed to be independently distributed over j and to have mean zero and finite variance. Moreover, suppose that we employ the general smoothing criterion

$$\text{PENSSE}_\lambda(\hat{x}) = n^{-1} \sum_j [y_j - \hat{x}(t_j)]^2 + \lambda \int (L\hat{x})^2(t)\, dt \tag{13.6}$$

for some differential operator L. Then it is not difficult to show (see Wahba, 1990) that, if we choose $\hat{x}$ to minimize PENSSE_λ, then the integrated squared bias

$$\text{Bias}^2(\hat{x}) = \int \{\text{E}\hat{x}(t) - x(t)\}^2\, dt$$

cannot exceed $\int Lx(t)^2\, dt$. This is rather useful, because if we choose L so that $Lx = 0$ even approximately, the bias is sure to be small.

As a consequence, we can use a relatively larger value of the smoothing parameter, leading to lower variance, without introducing excessive bias. Also, we can achieve a small value of the integrated mean squared error

$$\text{IMSE}(\hat{x}) = \int \text{E}[\hat{x}(t) - x(t)]^2\, dt$$

since

$$\text{IMSE}(\hat{x}) = \text{Bias}^2(\hat{x}) + \text{Var}(\hat{x})$$

where

$$\text{Var}(\hat{x}) = \int \text{E}\{\hat{x}(t) - \text{E}[\hat{x}(t)]\}^2\, dt.$$

Therefore we can conclude that, given any prior knowledge at all about the predominant shape of x, it is worth choosing a linear differential operator L to annihilate functions having that shape. In the next two chapters we show how to construct customized spline smoothers of this type .

This insight about the role of L in the regularization process also leads to the following interesting question: Can we use the information in N replications x_i of functional observations such as growth or temperature curves to *estimate* an operator L that comes close in some sense to satisfying $Lx_i = 0$? If so, then we should certainly use this information to improve on our smoothing techniques. This matter is taken up in Chapter 15.

13.3.3 Operators to partition variation

A central fact about linear differential operators L of the form (13.1) of degree m is that, in general, there are m linearly independent solutions

u_j of the homogeneous equation $Lu_j = 0$. There is no unique way of choosing these m functions u_j, but any choice is related by a linear transformation to any other choice. The set of all functions z for which $Lz = 0$ is called the *null space* of L, and the functions u_j form a basis for this space.

Consider, for example, the derivative operator $L = D^m$: The monomials $\{1, t, \dots t^{m-1}\}$ are a basis for the null space, as is the set of m polynomials formed by any nonsingular linear transformation of these. Likewise the functions $\{1, e^{\gamma t}\}$ are a solution set for $-\gamma Dx + D^2 x = 0$, and $\{1, \sin \omega t, \cos \omega t\}$ were cited as the solution set or null space functions for $Lx = \omega^2 Dx + D^3 x = 0$ in Chapter 1.

Thus, we can use linear differential operators L to partition functional variation in the sense that Lx splits x into two parts, the first consisting of whatever in x is a linear combination of the null space functions u_j, and the second being whatever is orthogonal to these functions. As we already know, this partitioning of variation is just what happens with basis functions ϕ_k and the projection operator P that expands x as a linear combination of these basis functions. The projection

$$Px = \hat{x} = \sum_{k=1}^{m} c_k \phi_k$$

splits any function x into the component $\hat{x}$ that is an optimal combination of the basis functions in a least squares sense, and an orthogonal residual component $x - \hat{x} = (I - P)x$. The complementary projection operator $Q = I - P$ therefore satisfies the linear homogeneous equation $Q\hat{x} = 0$, as well as the m equations $Q\phi_k = 0$. Thus the projection operator Q and the differential operator L appear to have analogous properties.

But there are some important differences. First, projection does not pay any attention to derivative information, whereas L does. Second, we have the closely related fact that Q is chosen to make Qx small, whereas L is chosen to make Lx small. Since Lx involves derivatives up to order m, making Lx small inevitably means paying attention to the size of $D^m x$. If we think there is important information in derivatives in some vague sense, it seems right to exploit this in splitting variation.

It is particularly easy to compare the two operators, differential and projection, in situations where there is an orthonormal basis expansion for the function space in question. Consider, for example, the space of infinitely differentiable periodic functions defined on the interval $[0, 1]$ that would be natural for modelling our temperature and precipitation records. A function x has the Fourier expansion

$$x(t) = c_0 + \sum_{k=1}^{\infty} [c_{2k-1} \sin(2\pi kt) + c_{2k} \cos(2\pi kt)].$$

Suppose our two operators L and Q are of order three and designed to eliminate the first three terms of the expansion, that is, a vertically shifted sinusoidal variation of period one. Then,

$$Qx(t) = \sum_{k=2}^{\infty} [c_{2k-1} \sin(2\pi kt) + c_{2k} \cos(2\pi kt)],$$

whereas

$$\begin{aligned} Lx &= 4\pi^2 Dx + D^3 x \\ &= \sum_{k=2}^{\infty} 8\pi^3 k(k^2 - 1)[-c_{2k-1} \cos(2\pi kt) + c_{2k} \sin(2\pi kt)]. \end{aligned}$$

Note that applying Q does not change the expansion beyond the third term, whereas L multiplies each successive pair of sines and cosines by an ever-increasing factor proportional to $k(k^2 - 1)$. Thus, L actually accentuates high-frequency variation whereas Q leaves it untouched; functions that are passed through L are going to come out rougher than those passing through Q.

The consequences for smoothing are especially important: If we penalize the size of $\|Lx\|^2$ in spline smoothing by minimizing the criterion (13.6), the roughening action of L means that high-frequency components are forced to be smaller than they would be in the original function, or than they would be if we penalized using Q by using the criterion

$$\text{PENSSE}_\lambda^Q(\hat{x}) = n^{-1} \sum_j [y_j - \hat{x}(t_j)]^2 + \lambda \int (Q\hat{x})^2(t)\, dt. \tag{13.7}$$

But customizing a regularization process is only one reason for splitting functional variation, and in Chapter 14 we look at a differential operator analogue of principal components analysis, called *principal differential analysis*, that can prove to be a valuable exploratory tool.

13.3.4 Operators to define solutions to problems

Engineers and scientists find that differential equations in general and linear differential equations in particular often provide elegant and powerful tools for identifying the solution of a problem or stating a theory. For example, Newton might have defined his Third Law as the solution of the equation

$$MD^2s = F$$

where s is the position of a body with mass M as a function of time.

For example, consider the simple but important problem of defining an arbitrary function that is everywhere positive and at the same time differentiable, with domain $t \geq 0$. One approach is to define it as the solution of the equation

$$Dx = wx$$

where the coefficient function w is smooth enough to be integrated. That a positive differentiable function satisfies this equation is clear; since x is positive and differentiable, we can take the ratio Dx/x, and this is simply w. Conversely, the general solution of the equation $Dx = wx$ is given by

$$x(t) = \beta \exp\left(\int_0^t w(u)du\right), \tag{13.8}$$

where β is an arbitrary positive constant. What we have achieved here is to change a constrained problem, namely defining a positive function x, into an unconstrained problem, that is computing w. In general, it is much easier to approximate unconstrained functions than constrained ones.

Taking this one step further, the equation

$$D^2x = wDx$$

defines an arbitrary twice-differentiable monotonic function, as in Section 5.4.2. As explained there, the solution to this equation is of the form

$$x(t) = x_0 + \alpha \int_0^t \exp\left(\int_0^u w(z)\,dz\right) du \tag{13.9}$$

where x_0 and α are arbitrary constants with $\alpha \neq 0$. This provides the general solution for a twice-differentiable monotonic function, and at the same time transforms the problem from estimating the constrained function x to estimating the unconstrained function w. Note, too, that this generalizes the equation for vertically-shifted exponential growth $x(t) = x_0 + \alpha e^{\beta t}$ in an interesting way; the constant function $w(t) = \beta$ yields this equation when substituted in (13.9), implying that departures from constant behaviour of w generalize exponential growth patterns in the function x.

13.4 Some linear differential equation facts

So far in this chapter, we have set the scene for the use of differential operators in FDA. Now we move on to a more detailed discussion of techniques and ideas that we use in this and the following chapters. Readers familiar with the theory of linear ordinary differential equations may wish to skip on to the next two chapters, and refer back to this material only where necessary. In any case, it is far beyond the scope of this book to offer even a cursory treatment of a topic as rich as the theory of differential equations, and there would be little point, since there are many fine texts on the topic. Some of our favourites that are also classics are Coddington (1989), Coddington and Levinson (1955) and Ince (1956), and for advice on a wide range of practical matters we recommend Press et al. (1992).

13.4.1 Derivatives are rougher

First, it is useful to point out a few things of general importance. For example, taking a derivative is generally a roughening operation, as we have observed in the context of periodic functions. This means that Dx in general has rather more curvature and variability than x. It is perhaps unfortunate that our intuitions about functions are shaped by our early exposure to polynomials, for which derivatives are smoother than the original functions, and transcendental functions such as e^t and $\sin t$, for which taking derivatives produces essentially no change in shape. In fact, the general situation is more like the growth curve accelerations in Figure 1.2, which are much more variable than the height curves in Figure 1.1, or the result of applying the third order LDO to temperature functions displayed in Figure 1.4.

By contrast, the operation of partial integration essentially reverses the process of differentiation (except for the constant of integration). It is convenient to use the notation $D^{-1}x$ for

$$D^{-1}x(t) = \int_{t_0}^{t} x(s)\,ds,$$

relying on context to specify the lower limit of integration t_0. This means, of course, that $D^{-1}Dx = x$. In fact, much of the theory of LDO's can be simplified, at least informally, by using the notion that various powers of D, such as D^3, D^{-2} and so on, behave essentially like polynomials, and can be factored in the same way. The partial integration operator D^{-1} is a smoothing operator in the sense that the result has less variability and curvature in general.

13.4.2 Finding a LDO that annihilates known functions

We have already cited a number of examples where we had a set of known functions $\{u_1, \dots, u_m\}$ and where at the same time we were aware of the LDO L that solved the homogeneous linear differential equations $Lu_j = 0, j = 1, \dots, m$. However, suppose that we have the u_j's in mind but that the LDO that annihilates them is not obvious, and we want to find it.

The process of identifying the LDO that sets m linearly independent functions to 0, as well as other aspects of working with LDO's, can be exhibited through the following example. Suppose we are considering an amplitude-modulated sinusoidal signal with fixed period ω. Such a signal would be of the form

$$x(t) = c_1 A(t) \sin(\omega t) + c_2 A(t) \cos(\omega t). \tag{13.10}$$

The function A determines the amplitude pattern. If A is regarded as a known time-varying function, the constants c_1 and c_2 determine the overall size of the signal's amplitude and its phase.

For given ω and $A(t)$, our aim is to find a differential operator L such that the null space of L consists of all functions of the form 13.10. Because these functions form a linear space of dimension two, we seek an annihilating operator of order two, of the form

$$Lx = w_0 x + w_1 Dx + D^2.$$

The task is to estimate the two weight functions w_0 and w_1.

First, let us do a few things to streamline the notation. Define the vector functions u and w as

$$u(t) = \begin{bmatrix} A(t)\sin(\omega t) \\ A(t)\cos(\omega t) \end{bmatrix} \text{ and } w(t) = \begin{bmatrix} w_0(t) \\ w_1(t) \end{bmatrix}. \tag{13.11}$$

Also, use $\mathsf{S}(t)$ to stand for $\sin(\omega t)$ and $\mathsf{C}(t)$ for $\cos(\omega t)$. Then

$$u = \begin{bmatrix} A\mathsf{S} \\ A\mathsf{C} \end{bmatrix}. \tag{13.12}$$

The required differential operator L satisfies the vector equation $Lu = 0$.

Recall that the first and second derivatives of S are $\omega\mathsf{C}$ and $-\omega^2\mathsf{S}$, respectively, and that those of C are $-\omega\mathsf{S}$ and $-\omega^2\mathsf{C}$, respectively. And, of course, the ever-important $\mathsf{S}^2 + \mathsf{C}^2 = 1$. Then the first two derivatives of u are, after a bit of simplification,

$$Du = \begin{bmatrix} (DA)\mathsf{S} + \omega A\mathsf{C} \\ (DA)\mathsf{C} - \omega A\mathsf{S} \end{bmatrix}$$

and

$$D^2u = \begin{bmatrix} (D^2A)\mathrm{S} + 2\omega(DA)\mathrm{C} - \omega^2 A\mathrm{S} \\ (D^2A)\mathrm{C} - 2\omega(DA)\mathrm{S} - \omega^2 A\mathrm{C} \end{bmatrix}. \tag{13.13}$$

By taking the second derivatives over to the other side of the equation, the relationship $Lu = 0$ can be expressed as follows:

$$w_0 u + w_1 Du = -D^2 u \tag{13.14}$$

or, in matrix notation,

$$\begin{bmatrix} u & Du \end{bmatrix} w = -D^2 u. \tag{13.15}$$

This is a linear matrix equation for the unknown weight functions w_0 and w_1, and its solution is simple provided that the matrix

$$\mathbf{W}(t) = \begin{bmatrix} u(t) & Du(t) \end{bmatrix} \tag{13.16}$$

is nowhere singular, or in other words that its determinant $|\mathbf{W}(t)|$ does not vanish for any value of the argument t. This coefficient matrix, which plays an important role in LDO theory, is called the *Wronskian matrix*, and its determinant is called the *Wronskian* for the system.

In this example, substituting the specific functions $A\mathrm{S}$ and $A\mathrm{C}$ for u_1 and u_2 gives

$$\mathbf{W} = \begin{bmatrix} A\mathrm{S} & (DA)\mathrm{S} + \omega A\mathrm{C} \\ A\mathrm{C} & (DA)\mathrm{C} - \omega A\mathrm{S} \end{bmatrix}. \tag{13.17}$$

Thus the Wronskian is

$$|\mathbf{W}| = A\mathrm{S}[(DA)\mathrm{C} - \omega A\mathrm{S}] - A\mathrm{C}[(DA)\mathrm{S} + \omega A\mathrm{C}] = -\omega A^2 \tag{13.18}$$

after some simplification. We have no worries about the singularity of $\mathbf{W}(t)$, then, so long as the amplitude function $A(t)$ does not vanish.

The solutions for the weight functions are then given by

$$w = -\mathbf{W}^{-1} D^2 u.$$

This takes a couple of sheets of paper to work out, or preferably is solved using symbolic computation software such as Maple (Char et al., 1991) or Mathematica (Wolfram, 1991). Considerable simplification is possible because of the identity $\mathrm{S}^2 + \mathrm{C}^2 = 1$, leading to

$$\mathbf{W} = \begin{bmatrix} \omega^2 + 2(DA/A)^2 - D^2A/A \\ -2DA/A \end{bmatrix}, \tag{13.19}$$

so that, for any function x,

$$Lx = \{\omega^2 + 2(DA/A)^2 - D^2A/A\}x - 2\{(DA)/A\}(Dx) + D^2x.$$

As we should expect, the weight coefficients in (13.19) are scale free in the sense that multiplying $A(t)$ by any constant does not change them.

Consider two simple possibilities for amplitude modulation functions. When $A(t)$ is a constant, both derivatives vanish, the operator reduces to $L = \omega^2 + D^2$ and $Lx = 0$ is the equation for simple harmonic motion. On the other hand, if $A(t) = e^{-\lambda t}$ so that the signal damps out exponentially with rate λ, then things simplify to

$$w = \begin{bmatrix} \omega^2 + \lambda^2 \\ 2\lambda \end{bmatrix} \quad \text{or} \quad Lx = (\omega^2 + \lambda^2) + 2\lambda Dx + D^2 x. \tag{13.20}$$

This is the equation for damped harmonic motion with a damping coefficient 2λ.

The example illustrates the following general principles. First, the order m Wronskian matrix

$$\mathbf{W}(t) = \begin{bmatrix} u(t) & Du(t) & \dots & D^{m-1}u(t) \end{bmatrix} \tag{13.21}$$

must be invertible, so that its determinant must not vanish over the range of t being considered. There are ways of dealing with isolated singularities, however, but these are beyond our present scope. Second, finding the vector of weight functions $w = (w_0(t), \dots, w_{m-1}(t))'$ is then a matter of solving the system of m linear equations $\mathbf{W}(t)w(t) = -D^m u(t)$, again with the possible aid of symbolic computation software.

13.4.3 *Finding the functions u_j satisfying $Lu_j = 0$*

Now let us consider the problem converse to that discussed in Section 13.4.2. Given an LDO L of order m, we might wish to identify m linearly independent solutions u_j to the homogeneous equation $Lx = 0$. We can do this directly by elementary calculus in simple cases, but more generally there is a variety of analytic and numerical approaches to this problem. For full details, see a standard reference on numerical methods, such as Stoer and Bulirsch (1980).

One particular numerical iterative procedure, Picard's algorithm, is simple to implement and, in principle, guaranteed to converge. Define the order m matrix function $\mathbf{T}$ as 0 for all values of t except for

$$\mathbf{T}_{j,j+1}(t) = 1, \ j = 1, \dots, m-1$$

and

$$\mathbf{T}_{mj}(t) = -w_{j-1}(t), \ j = 1, \dots, m.$$

Initialize the order m matrix function $\mathbf{U}$ as $\mathbf{U}^{(0)}(t) = \mathbf{I}$. On iteration $\nu \geq 1$, $\mathbf{U}^{(\nu)}$ is updated as follows:

$$\mathbf{U}^{(\nu)}(t) = \mathbf{I} + \int_0^t \mathbf{T}(s)\mathbf{U}^{(\nu-1)}(s)\, ds. \tag{13.22}$$

If we plot the results of each iteration, we discover that they converge moving from left to right; on a particular iteration ν, the u_j have converged up to some argument $t < T$, whereas the remainder or tails of the functions are still varying, quite often wildly, from one iteration to the next. This phenomenon is helpful in determining how long one has to proceed to get total convergence, and can also be useful in speeding up the algorithm.

Upon convergence, $\mathbf{U}$ contains the solution functions u_j in the first row, and their successive derivatives in subsequent rows. We can carry out the integration in (13.22) using the simple trapezoidal rule, providing that the weight functions are available on a sufficiently fine grid of argument values.

13.5 Constraint functionals B

13.5.1 Some constraint examples

We have already noted that, in general, the space of solutions of the linear differential equation $Lx = 0$ is a function space of dimension m, called the null space of L. We shall assume that the linearly independent functions $u_1, \ldots, u_m$ form a basis of the null space.

Any specific solution of $Lx = 0$ then requires m additional pieces of information about x. For example, we can solve the equation $-\gamma Dx + D^2x = 0$, defining a shifted exponential, exactly provided that we are able to specify that

$$x(0) = 0 \text{ and } Dx(0) = 1$$

in which case

$$x(t) = \gamma^{-1}e^{\gamma t} - 1.$$

Alternatively, $x(0) = 1$ and $Dx(0) = 0$ imply that $x_0 = 1$ and $\alpha = 0$ in (13.3), or simply that $x = 1$.

We introduce the notion of a *constraint operator B* to specify the m pieces of information about x that we require to identify a specific function x as the unique solution to $Lx = 0$. This operator simply *evaluates* x or its derivatives in m different ways. The most important example is the *initial value* operator used in the theory of ordinary differential equations defined over an interval $\mathcal{T} = [0, T]$,

$$\textbf{Initial Operator: } B_I x = \begin{bmatrix} x(0) \\ Dx(0) \\ \vdots \\ D^{m-1}x(0) \end{bmatrix}. \tag{13.23}$$

When $B_I x$ is set to an m-vector, initial value constraints are defined. In the example above, we considered the two cases $B_I x = (0, 1)'$ and $B_I x = (1, 0)'$, implying the two solutions given there.

The following *boundary value* operator is also of great importance in applications involving LDO's of even degree:

$$\textbf{Boundary Operator: } B_B x = \begin{bmatrix} x(0) \\ x(T) \\ \vdots \\ D^{(m-2)/2}x(0) \\ D^{(m-2)/2}x(T) \end{bmatrix}. \tag{13.24}$$

Specifying $B_B x = c$ gives the values of x and its first $(m - 2)/2$ derivatives at both ends of the interval of interest.

The *periodic constraint* operator is defined by

$$\textbf{Periodic Operator: } B_P x = \begin{bmatrix} x(T) - x(0) \\ Dx(T) - Dx(0) \\ \vdots \\ D^{m-1}x(T) - D^{m-1}x(0) \end{bmatrix}. \tag{13.25}$$

Functions satisfying $B_P x = 0$ are periodic up to the derivative D^{m-1} over $\mathcal{T}$, and are said to obey periodic boundary conditions.

13.5.2 How L and B partition functions

Whatever constraint operator we use, consider the problem of expressing any particular function x as a sum of two components z and e, such that $Lz = 0$ and $Be = 0$. When can we carry out this partitioning in a unique way? This happens if and only if $x = 0$ is the only function satisfying both $Bx = 0$ and $Lx = 0$, or, in algebraic notation,

$$\ker B \cap \ker L = 0. \tag{13.26}$$

Thus, the two operators B and L complement each other; the equation $Lx = 0$ defines a space of functions $\ker L$ that is of dimension m, and within this space B is a nonsingular transformation. Looking at it the other way, the equation $Bx = 0$ defines a space of functions $\ker B$ of *co-dimension* m, within which L is a one-to-one transformation.

Note that the condition (13.26) can break down. For example, consider the operator $L = \omega^2 I + D^2$ on the interval $[0, T]$. The space $\ker L$ contains all linear combinations of $\sin \omega t$ and $\cos \omega t$. If $\omega = 2\pi k/T$ for some integer k and we use boundary constraints, all multiples of $\sin \omega t$ satisfy $B_B x = 0$, and so the condition (13.26) is violated. In this case, some functions—those that satisfy $x(0) = x(T)$ and $Dx(0) = Dx(T)$—has infinitely many decompositions as $z + e$ with $Lz = Be = 0$, wheras we will not be able to decompose others at all in this way.

If we place constraints on the functions under consideration, periodic boundary conditions for example, then the dimension of $\ker L$ may be less than m, and a lower dimensional constraint operator is appropriate. For example, consider the operator $L = \omega^4 I - D^4$ on the interval $[0, T]$, with T a multiple of $2\pi/\omega$. Then the space of functions satisfying $Lu = 0$ and periodic boundary conditions is 2, and is spanned by the functions $\sin \omega t$ and $\cos \omega t$. To specify a function completely, we would most simply give the values $x(0)$ and $Dx(0)$, so an appropriate operator B is defined by $Bx = (x(0), Dx(0))'$. Then the condition (13.26) is satisfied.

13.5.3 *The inner product defined by operators L and B*

All the functional data analysis techniques and tools in this book depend on the notion of an inner product between two functions x and y. We have seen numerous examples where a careful choice of inner product can produce useful results, especially in controlling the roughness of estimated functions, such as functional principal components or regression functions. In these and other examples, it is important to use derivative information in defining an inner product.

Let us assume that the constraint operator is such that condition (13.26) is satisfied. We can define a large family of inner products as follows:

$$\langle x, y \rangle_{B,L} = (Bx)'(By) + \int (Lx)(t)(Ly)(t)\, dt \tag{13.27}$$

with the corresponding norm

$$\|x\|^2_{B,L} = (Bx)'(Bx) + \int (Lx)^2(t)\, dt. \tag{13.28}$$

The condition (13.26) ensures that this is a norm; the only function x for which $\|x\|_{B,L} = 0$ is zero itself, since this is the only function simultaneously satisfying $Bx = 0$ and $Lx = 0$.

In fact, this inner product works by splitting the function x into two parts:

$$x = z + e, \text{ where } z \in \ker L \text{ and } e \in \ker B.$$

The first term in (13.28) simply measures the size of the component z, since $Be = 0$ and therefore $Bx = Bz$, whereas the second term depends only on the size of the component e since $Lx = Le$. The first term in (13.27) is essentially an inner product for the m-dimensional subspace in which z lives and which is defined by $Lz = 0$. The second term is an inner product for the function space of codimension m defined by $Be = 0$. Thus, we can write

$$\|x\|_{B,L}^2 = \|z\|_B^2 + \|e\|_L^2.$$

With this composite inner product in hand, that is with a particular LDO L and constraint operator B in mind, we can go back and revisit each of our functional data analytic techniques to see how they perform with this inner product. This is the central point explored by Ramsay and Dalzell (1991), to which we refer the reader for further discussion.

13.6 The Green's function for L and B

13.6.1 Solving a linear differential equation

Suppose that we want to reverse the effect of applying an mth order LDO L, that is, we have a *forcing function* f satisfying

$$Lx = f \tag{13.29}$$

and we want to find x. We can recognise that the solution is not unique; if we add any linear combination of the functions u_j that span $\ker L$ to a solution x, then this function also satisfies the equation. Throughout this section, assume that the constraint operator B satisfies (13.26). Then, for any given vector c, the additional constraints

$$Bx = c \tag{13.30}$$

uniquely specify x. Can we be assured that a solution x exists at all? If we are using initial value constraints, then under mild regularity conditions on f, there is indeed a solution.

Let us assume, as is the case for initial value constraints, that the space $\ker L$ is of dimension m. Define the matrix $\mathbf{B}$ in vector notation by

$$\mathbf{B} = Bu' \text{ so that } \mathbf{B}_{ij} = (Bu_j)_i. \tag{13.31}$$

Every z in $\ker L$ can be written as $\sum_j a_j u_j$ for an m-vector of coefficients a, and $Bz = \mathbf{B}a$ by the definition of $\mathbf{B}$. The conditions we have specified ensure that $\mathbf{B}$ is invertible. In passing we note that if we find the functions u_j by Picard's algorithm as outlined in Section 13.4.3, then $\mathbf{B}$ is the identity.

Returning to our constrained problem, set the vector $a = \mathbf{B}^{-1}c$ and let $z = \sum_j a_j u_j$. Suppose e satisfies $Le = f$ subject to $Be = 0$; then $x = z + e$ satisfies $Lx = f$ subject to $Bx = c$. Therefore, if we can solve the problem

$$Lx = f \text{ subject to } x \text{ in } \ker B\,, \tag{13.32}$$

we can find a solution subject to the more general constraint $Bx = c$.

13.6.2 The Green's function and the solution

It can be shown that there exists a bivariate function $G(t;s)$, called the *Green's function* associated with the pair (B, L), such that

$$x(t) = \int_T G(t;s)Lx(s)\,ds \text{ for } x \text{ in } \ker B. \tag{13.33}$$

Thus the Green's function defines an integral transform $\mathcal{G} = \int G(t;\cdot)$ that inverts the linear differential operator L. Applying $\mathcal{G}$ to Lx gets us back to x itself, provided $Bx = 0$.

Furthermore, under mild regularity conditions, $\mathcal{G}$ is the inverse of L in the other sense: If we apply $\mathcal{G}$ to a function f and then L to the result, we recover the function f. This means that if we set $x = \mathcal{G}f$, then

$$x(t) = \int_T G(t;s)f(s)\,ds,$$

and we have solved the equation $Lx = f$ subject to $Bx = 0$, which is what we set out to do.

Before giving a general recipe for computing the Green's function G, let us look at a specific example. If our interval is $[0, T]$ and our constraint operator is the initial value constraint $B_I x = x(0)$, then for $L = D$,

$$G_I(t;s) = 1, s \le t, \text{ and } 0 \text{ otherwise.}$$

where the subscript on G is to remind us that this applies only for the initial value constraint operator B_I.

This follows because

$$\int_0^T G_I(t;s)Dx(s)\,ds = \int_0^t Dx(s)\,ds = x(t) - x(0).$$

Because of the initial value constraint $x(0) = 0$, we finally have the desired result (13.33).

13.6.3 *A recipe for the Green's function*

We can now offer a recipe for constructing the Green's function for any LDO L of the form (13.1) and the initial value constraint B_I of the corresponding order. First, compute the Wronskian matrix $\mathbf{W}$ defined in (13.21). Second, define the functions $v_1, \ldots, v_m$ to be the elements of the last row of $\mathbf{W}^{-1}$. Then, it turns out that

$$G_I(t;s) = \sum_{j=1}^{m} u_j(t)v_j(s) = u(t)'v(s), s \le t, \text{ and 0 otherwise} \quad (13.34)$$

where v is the vector-valued function with components v_j.

Let us see how this works for $L = -\gamma D + D^2$. The space $\ker L$ is spanned by the two functions $u_1(t) = 1$ and $u_2(t) = \exp \gamma t$. The Wronskian matrix is

$$\mathbf{W}(t) = \begin{bmatrix} u_1(t) & Du_1(t) \\ u_2(t) & Du_2(t) \end{bmatrix} = \begin{bmatrix} 1 & 0 \\ \exp \gamma t & \gamma \exp \gamma t \end{bmatrix}$$

and consequently

$$\mathbf{W}^{-1}(t) = \begin{bmatrix} 1 & 0 \\ -\gamma^{-1} & \gamma^{-1} \exp -\gamma t \end{bmatrix}$$

from which

$$v(s) = \gamma^{-1}(-1, \exp -\gamma s)'$$

and finally

$$G_I(t;s) = \gamma^{-1}(e^{\gamma(t-s)} - 1), s \le t, \text{ and 0 otherwise.} \quad (13.35)$$

We can verify that this is the required Green's function by integrating by parts.

We do not discuss in any detail the case of any constraint functions B other than initial value constraints. Under boundary or periodic constraints, it may be that additional conditions are required on the function f or on the constraint values c, but nevertheless we can extend the basic ideas of Green's functions. For details, we refer the reader to Coddington and Levinson (1955), Roach (1982) or most other texts on linear differential equations. For more details along the lines we present in the following section, see Dalzell and Ramsay (1990).

13.7 Reproducing kernels and Green's functions

A bivariate function called the *reproducing kernel* plays a central role in the theory of spline functions. There are two to consider, one for each of the function subspaces $\ker B$ and $\ker L$. In the context of these two subspaces, we shall explain what a reproducing kernel is, and show how it can be constructed in these specific cases.

13.7.1 *The reproducing kernel for* $\ker B$

Consider, first, the subspace $\ker B$ consisting of functions that satisfy $Bx = 0$. The reproducing kernel for this subspace has a simple relationship to the Green's function G.

Given any two functions x and y in $\ker B$, let us define the L-inner product by

$$\langle x, y\rangle_L = \langle Lx, Ly\rangle = \int Lx(s)Ly(s)\,ds.$$

Let G_I be the Green's function as defined in Section 13.6.3, and define a function $k_2(t,s)$ such that, for all t,

$$Lk_2(t,\cdot) = G_I(t;\cdot) \text{ and } Bk_2(t,\cdot) = 0. \tag{13.36}$$

By the defining properties of Green's functions, this means that

$$k_2(t,s) = \int G_I(s;w)G_I(t;w)\,dw. \tag{13.37}$$

The function k_2 has an interesting property. Suppose that e is any function in $\ker B$, and consider the L-inner product of $k_2(t,\cdot)$ and e. For all t,

$$\langle k_2(t,\cdot), e\rangle_L = \int Lk_2(t,s)Le(s)\,ds = \int G_I(t;s)Le(s)\,ds = e(t) \tag{13.38}$$

by the key property (13.33) of Green's functions. Thus, in the space $\ker B$, taking the L-inner product of k_2 regarded as a function of its second argument with any function e yields the value of e at the first argument of k_2. Thus, taking the inner product with k_2 reproduces the function e, and k_2 is called the reproducing kernel for this function space and inner product.

Chapter 15 shows that the reproducing kernel is the key to the important question, "Is there an optimal set of basis functions for smoothing data?" To answer this question, we need to use the important property

$$\langle k_2(s,\cdot), k_2(t,\cdot)\rangle_L = k_2(s,t), \tag{13.39}$$

which follows at once from (13.38) setting $e(\cdot) = k_2(s, \cdot)$ and appealing to the symmetry of the inner product.

We can put the expression (13.37) in a slightly more convenient form for the purpose of calculation. Recall the definitions of the vector-valued functions u and v in Section 13.6.3, and define the order m symmetric matrix-valued function $\mathbf{F}(s)$ as

$$\mathbf{F}(s) = \int_0^s v(w)v(w)'\,dw. \tag{13.40}$$

Then, assuming that $s \le t$, the formula (13.34) gives

$$k_2(s,t) = \int_0^s [u(s)'v(w)][v(w)'u(t)]\,dw = u(s)'\mathbf{F}(s)u(t) \tag{13.41}$$

To deal with the case $s > t$, we use the property that $k_2(s,t) = k_2(t,s)$.

13.7.2 The reproducing kernel for ker L

Now suppose that $f = \sum a_i u_i$ and $g = \sum b_i u_i$ are elements of $\ker L$. We can consider the B-inner product on the finite-dimensional space $\ker L$, defined by

$$\langle f, g\rangle_B = (Bf)'Bg = a'\mathbf{B}'\mathbf{B}b.$$

Define a function $k_1(t,s)$ by

$$k_1(t,s) = u(t)'(\mathbf{B}'\mathbf{B})^{-1}u(s).$$

Now it is easy to verify that, for any $f = \sum_i a_i u_i$,

$$\langle k_1(t,\cdot), f\rangle_B = u(t)'(\mathbf{B}'\mathbf{B})^{-1}\mathbf{B}'\mathbf{B}a = u(t)'a = a'u(t) = f(t).$$

So k_1 is the reproducing kernel for the space $\ker L$ equipped with the B-inner product.

Finally, we consider the space of more general functions x equipped with the inner product $\langle\cdot,\cdot\rangle_{B,L}$ as defined in Section 13.5.3. It is easy to check from the properties we have set out that the reproducing kernel in this space is given by

$$k(s,t) = k_1(s,t) + k_2(s,t).$$

14
Principal differential analysis

14.1 Introduction

14.1.1 *The basic principle*

In this chapter we return to the problem that animated the development of principal components analysis in Chapter 6: can we use our set of N functional observations x_i to define a much smaller set of m functions u_j on the basis of which we can obtain efficient approximations of these observed functions?

What changes here, however, is what we mean by efficient approximation. In this chapter we consider the identification of a linear differential operator

$$L = w_0 I + w_1 D + \ldots + w_{m-1} D^{m-1} + D^m \tag{14.1}$$

that comes as close as possible to satisfying the homogeneous linear differential equation $Lx_i = 0$ for each observation x_i. Thus, we seek a differential equation model so that our data satisfy

$$D^m x_i = -w_0 x_i - w_1 D x_i - \ldots - w_{m-1} D^{m-1} x_i$$

to the best possible degree of approximation.

The term *principal differential analysis* (abbreviated PDA) is used for this methodology. This term was introduced in the paper Ramsay (1996a), on which the treatment of the current chapter is based. The

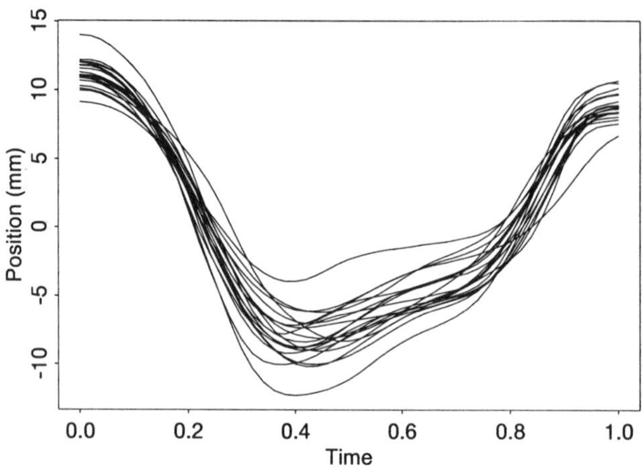

FIGURE 14.1. Twenty records of the position of the centre of the lower lip during the uttering of the syllable "bob." These curves are the result of preliminary smoothing registration steps, and the time unit is arbitrary.

immediately obvious way of carrying out PDA is to adopt a least squares approach to the fitting of the differential equation model. Since we wish the operator L to annihilate the given data functions x_i as nearly as possible, we regard the function Lx_i as the residual error from the fit provided by the linear differential operator L. As the fitting criterion, the least squares approach then gives the sum of squared norms

$$\text{SSE}_{\text{PDA}}(L) = \sum_{i=1}^{N} \int [Lx_i(t)]^2 \, dt, \tag{14.2}$$

which can be minimized over L to find an estimate of the appropriate LDO. Of course, the operator L is defined by the m-vector w of functions $(w_0, \ldots, w_{m-1})'$, and so $\text{SSE}_{\text{PDA}}(L)$ depends on these functions, and estimating L is equivalent to estimating the m weight functions w_i.

14.1.2 *The need for PDA*

There are several reasons why a PDA can provide important information about the data and the phenomenon under study. Primarily, in many applications the differential equation $Lx = 0$ offers an interesting and useful way of understanding the processes that generated the data. As an example to be used throughout this chapter, consider the curves presented in Figure 14.1. These indicate the movement of the centre of the lower lip as a single speaker said the syllable "bob". The time interval has been arbitrarily set to [0,1], although the actual syllable took about a third of a second. The displayed curves are the result of considerable preprocessing, including smoothing and the use of functional PCA to identify the direction in which most of the motion was found. Details can be found in Ramsay, Munhall, Gracco and Ostry (1996). In broad terms, we see that the lower lip motion has three phases: an initial rapid opening, a sharp transition to a relatively slow and nearly linear motion, and a final rapid closure.

Because the lower lip is part of a mechanical system, with certain natural resonating frequencies and a stiffness or resistance to movement, it seems appropriate to explore to what extent this motion can be expressed in terms of a second order linear differential equation of the type useful in analysing such systems,

$$Lx_i = w_0 x_i + w_1 D x_i + D^2 x_i = 0. \tag{14.3}$$

Discussions of second-order mechanical systems can be found in most applied texts on ordinary differential equations, such as Tenenbaum and Pollard (1963).

The first coefficient w_0 essentially reflects the position-dependent force applied to the system at position x. Coefficient values $w_0 > 0$ and $w_1 = 0$ correspond to a system with sinusoidal or harmonic motion, with $w_0^{1/2}/(2\pi)$ cycles per unit time and wavelength or period $2\pi w_0^{-1/2}$; w_0 is often called the *spring constant*. The second coefficient w_1 indicates influences on the system that are proportional to velocity rather than position and are often internal or external frictional forces or viscosity in mechanical systems.

The *discriminant* is defined as $d = (w_1/2)^2 - w_0$ and is critical in terms of its sign. When the discriminant is negative, the system is underdamped, and it exhibits oscillation. When d is positive, the system is overdamped, and either decays to zero or grows exponentially, in either case without oscillation. A critically damped system is one for which $d = 0$, and also exhibits nonoscillatory motion. When does the system exhibit exponential growth or instability? Always if $w_0 < 0$, and when $w_1 < 0$ in the event that $w_0 > 0$. When $w_0 = w_1 = 0$ the system is in linear motion, corresponding to $D^2 x = 0$.

Strictly speaking, these mechanical interpretations of the roles of the coefficient functions w_0 and w_1 are appropriate only if these functions are constants, but higher-order effects can be ignored if they do not

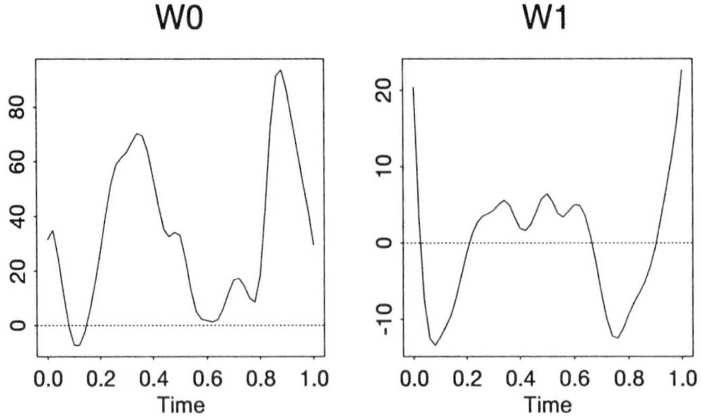

FIGURE 14.2. The two weight functions w_0 and w_1 estimated from the lip movement data.

vary too rapidly with t. In that case the signs of w_0, w_1, and d can be viewed as describing the instantaneous state of the system.

The techniques we develop were used to obtain the weight functions displayed in Figure 14.2. These are of rather limited help in interpreting the system, but one does note that w_0 is positive except for a brief episode near the beginning, and near zero in the central portion corresponding to the near-linear phase of lip movement. The two solutions to the homogeneous differential equation $Lu = 0$ defined by these weight functions are shown in Figure 14.3.

14.1.3 *The relationship to PCA*

Once we have found the operator L, in general we can define m linearly independent functions $u_1, \ldots, u_m$ that span the null space of L. Any function x that satisfies $Lx = 0$ can be expressed as a linear combination of the u_j. Hence, since L has been chosen to make the Lx_i as small as possible, we would expect to obtain a good approximation of the x_i by expanding them in terms of the u_j. This is closely reminiscent of PCA, where the first m principal component functions ξ_j form a good m-dimensional set for approximating the given data. The spirit of the approximation is rather different, however.

We can pursue the comparison between PCA and PDA by noting that PCA can also be considered to involve the identification of a linear operator, which we can denote by Q, such that the equation $Qx_i = 0$ is solved as nearly as possible. To see this, recall from Chapter 6 that one method of defining functional PCA is as the determination of a set

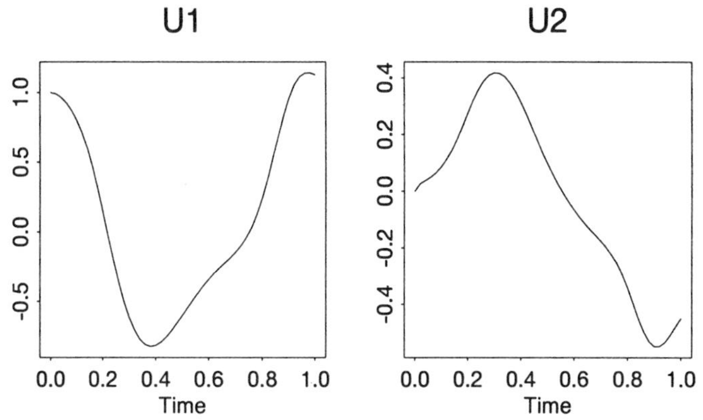

FIGURE 14.3. Two solutions of the homogeneous differential equation $Lu = 0$ estimated for the lip movement data.

of m basis functions ξ_j to minimize the least squares criterion

$$\mathrm{SSE}_{\mathrm{PCA}} = \sum_{i=1}^{N} \int [x_i(t) - \sum_{j=1}^{m} f_{ij}\xi_j(t)]^2 \, dt \tag{14.4}$$

with respect both to the basis functions ξ_j and with respect to the coefficients f_{ij} of the expansions of each observed function in terms of these basis functions. Because the minimization of the fitting criterion (14.4) is a least squares problem, we can think of PCA as a two-stage process: first identify a set of m orthonormal basis functions ξ_j, and then approximate any specific curve x_i by $\hat{x}_i = \sum_{j=1}^{m} f_{ij}\xi_j$. This second basis expansion step, the projection of each of the observed functions onto the m-dimensional space spanned by the basis functions ξ, takes place after determining the optimal basis for these expansions. Thus $\hat{x}_i$ is the image of x_i resulting from applying a least squares fit.

Suppose we indicate this projection as P_ξ, with the subscript indicating that the nature of the projection depends on the basis functions ξ_j, so that $P_\xi x_i = \hat{x}_i$. Associated with the projection P_ξ is the complementary projection

$$Q_\xi = I - P_\xi$$

which produces as its result the residuals

$$Q_\xi x_i = x_i - P_\xi x_i = x_i - \hat{x}_i.$$

Using this concept, we can alternatively and equivalently define the PCA problem in a way that is much more analogous to the problem of identifying the linear differential operator L: in PCA, one seeks a projection operator Q_ξ such that the residual sum of squares

$$\mathsf{SSE}_{\text{PCA}} = \sum_{i=1}^{N} \int [Q_\xi x_i(t)]^2 \, dt \tag{14.5}$$

is minimized. Indeed, one might think of the first m eigenfunctions as the functional *parameters* defining the projection operator Q_ξ, just as the weight functions w are the functional parameters of the LDO L in PDA. These eigenfunctions, and any linear combinations of them, exactly satisfy the equation $Q_\xi \xi_j = 0$, just as the m functions u_j referred to above exactly satisfy the equation $L_w u_j = 0$, where we now add the subscript w to L to remind ourselves that L is defined by the vector w of m weight functions w_j.

Principal differential analysis is defined, therefore, as the identification of the operator L_w that minimizes the least squares criterion $\mathsf{SSE}_{\text{PDA}}$, just as we can define PCA as the identification the projection operator Q_ξ that minimizes the least squares criterion $\mathsf{SSE}_{\text{PCA}}$.

Since the basic structures of the least squares criteria (14.2) and (14.5) are the same, clearly the only difference between the two criteria is in terms of the actions represented by the two operators L_w and Q_ξ. Since $Q_\xi x$ is in the same vector space as x, the definition of the operator identification problem as the minimization of $\|Q_\xi x\|^2$ is also in the same space, in the sense that we measure the performance of Q_ξ in the same space as the functions x to which it is applied.

On the other hand, L_w is a roughening transform in the sense that $L_w x$ has m fewer derivatives than x and is usually more variable. We may want to penalize or otherwise manipulate x at this rough level. Put another way, it may be plausible to conjecture that the noise or unwanted variational component in x is found only at the rough level $L_w x$. Thus, a second motivating factor for the use of L_w rather than Q_ξ is that the PDA process explicitly takes account of the smoothness of the data by first roughening the data before minimizing error, while PCA does not.

As an example, imagine that we are analysing the trajectories x_i of several rockets of the same type launched successively from some site. We observe that not all trajectories are identical, and we conjecture that some random process is at work that contributes variability to our observations. Naïvely, we might look for that variability in the trajectories themselves, but our friends in physics will be quick to point out firstly that the major source of variability is probably in the propulsion system, and secondly since the force that it applies is proportional to acceleration, we ought to study the acceleration

D^2x_i instead. Thus, if the function x_i is the trajectory along a specific coordinate axis (straight up, for example), the systematic or errorless trajectory should obey the law

$$f_i(t) = M(t)D^2x_i(t),$$

where $M(t)$ is the mass of the rocket at time t. Alternatively,

$$-f_i/M + D^2x_i = 0.$$

Taking a more empirical approach, however, we agree on the compromise of looking for a second-order linear differential equation

$$Lx = w_0x + w_1Dx + D^2x$$

and, if our friends in physics are right, the systematic or errorless component in the data should yield

$$w_0x_i = -f_i/M \text{ and } w_1 = 0.$$

What we do understand, in any case, is that the sources of variability are likely to be at the rough level D^2x_i, rather than at the raw trajectory level x_i.

Returning to the lip position curves, we might reason that variation in lip position from curve to curve is due to variation in the forces resulting from muscle contraction, and that these forces have a direct or proportional impact on the acceleration of the lip tissue, and thus only indirectly on the position itself. In short, position is two derivatives away from the action.

More generally, an important motivation for finding the operator L_w is substantive: applications in the physical sciences, engineering, biology and elsewhere often make extensive use of differential equation models of the form

$$Lx_i = f_i.$$

The result f_i is often called a *forcing or impulse function*, and in physical science and engineering applications is often taken to indicate the influence of exogenous agents on the system defined by $Lx = 0$.

Section 14.2 presents techniques for principal differential analysis, along with some measures of fit to the data. We also take up the possibility of regularizing or smoothing the estimated weight functions w_j.

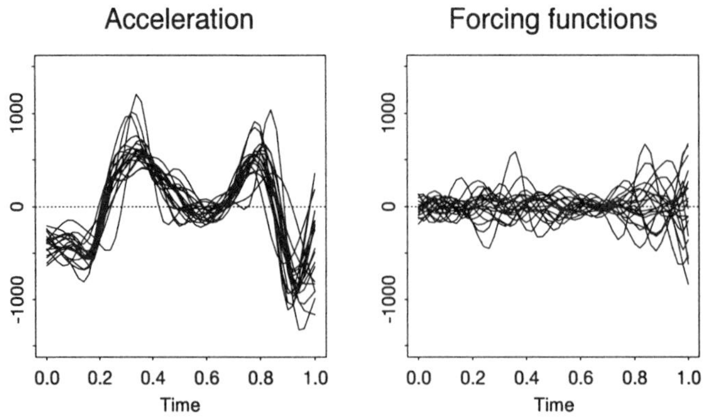

FIGURE 14.4. The left panel displays the acceleration curves $D^2 x_i$ for the lip position data, and the right panel the forcing functions Lx_i.

14.1.4 *Visualizing results*

One way to get a visual impression of how effective the differential operator L is at annihilating variation in the x_i is to plot the *empirical forcing functions* Lx_i. If these are small and mainly noise-like, we can have some confidence that the equation represents the data well. As a point of comparison, we can use the size of the $D^m x_i$'s, since these are the empirical forcing functions corresponding to $w_0 = \ldots = w_{m-1} = 0$. Figure 14.4 shows the acceleration curves for the lip data in the left panel, and the empirical forcing functions in the right. We see that the forcing functions corresponding to L are indeed much smaller in magnitude, and more or less noise-like except for two bursts of signal near the beginning and end of the time interval.

The value of the criterion SSE_{PDA} defined above is 6.06×10^6, whereas the same measure of the size of the $D^2 x_i$ is 40.67×10^6. If we call the latter measure SSY_{PDA}, then we can also summarize these results in the squared correlation measure

$$\text{RSQ}_{\text{PDA}} = (\text{SSY}_{\text{PDA}} - \text{SSE}_{\text{PDA}})/\text{SSY}_{\text{PDA}} \tag{14.6}$$

whose value is 0.85 for this problem.

Although, strictly speaking, it is not the task of PDA to approximate the original curves (this would be a job for PCA), we can nevertheless wonder how well the two solution curves would serve this purpose. Figure 14.5 shows the least squares approximation of the first two curves in terms of the two solution functions in Figure 14.3, and we see that the fit is fairly satisfactory.

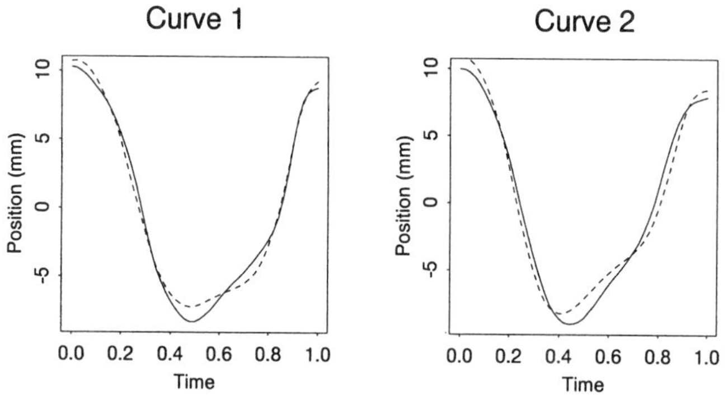

FIGURE 14.5. The solid curves are the first two observed lip position functions, and the dashed lines are their approximations on the basis of the two solution functions in Figure 14.3.

Finally, we return to the discriminant function $d = (w_1/2)^2 - w_0$, presented in Figure 14.6, and its interpretation. This system is more or less critically damped over the interval $0.5 \leq t \leq 0.75$, suggesting that its behaviour may be under external control. But in the vicinities of $t = 0.25$ and $t = 0.85$, the system is substantially underdamped, and thus behaves locally like a spring. The period of the spring would be around 30 to 40 msec, and this is in the range of values estimated in studies of the mechanical properties of flaccid soft tissue. These results suggest that the external input to lip motions tends to be concentrated in the brief period near $t = 0.6$, when the natural tendency for the lip to close is retarded to allow the articulation of the vowel.

14.2 PDA techniques

Now we turn to two techniques for estimating the weight functions w_j defining the linear differential operator that comes closest to annihilating the observed functions in the sense of criterion (14.2).

Once the operator L has been computed by estimating its weight functions w_j by one of these methods, we may be interested in computing a set of m linearly independent basis functions u_j satisfying $Lu_j = 0$. A variety of numerical techniques can be found in standard references on numerical methods, such as Stoer and Bulirsch (1980). The successive approximation or Picard's method, described in Chapter 13, works well in most situations.

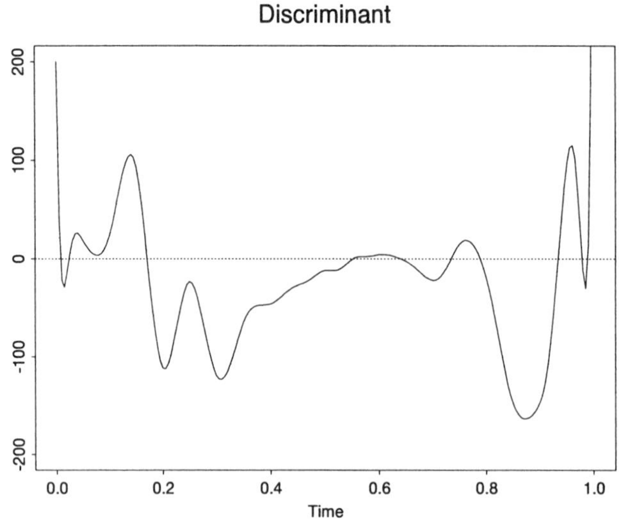

FIGURE 14.6. The discriminant function $d = (w_1/2)^2 - w_0$ for the second order differential equation describing lip position.

14.2.1 PDA by pointwise minimization

The first approach yields a pointwise estimate of the weight functions w_j computable by standard least squares estimation. Define the pointwise fitting criterion

$$\mathrm{PSSE}_L(t) = N^{-1} \sum_i (Lx_i)^2(t) = N^{-1} \sum_i [\sum_{j=0}^{m} w_j(t)(D^j x_i)(t)]^2 \quad (14.7)$$

where, as above, $w_m(t) = 1$ for all t. If t is regarded as fixed, the following argument shows that this is simply a least squares fitting criterion.

First define the m-dimensional coefficient vector as

$$w(t) = (w_0(t), \ldots, w_{m-1}(t))',$$

the $N \times m$ pointwise design matrix as

$$\mathbf{Z}(t) = \{(D^j x_i)(t)\}_{i=1,\ldots,N; j=0,\ldots,m-1}$$

and the N-dimensional dependent variable vector as

$$y(t) = \{-(D^m x_i)(t)\}_{i=1,\ldots,N}.$$

We can express the fitting criterion (14.7) in matrix terms as

$$\mathrm{PSSE}_L(t) = N^{-1}[y(t) - \mathbf{Z}(t)w(t)]'[y(t) - \mathbf{Z}(t)w(t)].$$

Then, holding t fixed, the least squares solution minimizing $\mathrm{PSSE}_L(t)$ with respect to the values $w_j(t)$ is

$$w(t) = [\mathbf{Z}(t)'\mathbf{Z}(t)]^{-1}\mathbf{Z}(t)'y(t). \tag{14.8}$$

The existence of these pointwise values $w(t)$ depends on the condition that the determinant of $\mathbf{Z}(t)'\mathbf{Z}(t)$ is bounded away from zero for all values of t, and it is wise to compute and display this determinant as a routine part of the computation. Assuming that the determinant is nonzero is equivalent to assuming that $\mathbf{Z}(t)$ is of full column rank for all t.

Of course, if m is not large, then we can express the solution in closed form. For example, for $m = 1$,

$$w_0(t) = -\sum_i x_i(t)(Dx_i)(t) \Big/ \sum_i x_i^2(t) \tag{14.9}$$

and the full-rank condition merely requires that, for each value of t, some $x_i(t)$ be nonzero.

Some brief comments about the connections with Section 13.4 are in order. There, we were concerned with finding a linear operator of order m that annihilated a set of exactly m functions u_i. For this to be possible, an important condition was the nonsingularity of the Wronskian matrix values $\mathbf{W}(t)$ whose elements were $D^j u_i(t)$. We obtain the matrix $\mathbf{Z}(t)$ from the functions x_i in the same way, but it is no longer a square matrix, because $N > m$ in general. However, the condition that $\mathbf{Z}(t)$ is of full column rank is entirely analogous.

14.2.2 PDA by basis expansions

The pointwise approach can pose problems in some applications. First, solving the equation $Lu = 0$ requires that the w_j be available at a fine level of detail, with the required resolution depending on their smoothness. Whether or not these functions are smooth depends in turn on the smoothness of the derivatives $D^j x_i$. Since we often estimate these derivatives by smoothing procedures that may not always yield smooth estimates for higher order derivatives, the resolution we require may be very fine indeed. But for larger orders m, computing the functions w_j pointwise at a fine resolution level can be computationally intensive, since we must solve a linear equation for every value of t for which w is required. This suggests the need for an approximate solution which can be computed quickly and which is reasonably regular or smooth.

Second, it may be desirable to circumvent the restriction that the rank of **Z** be full, especially if the failure is highly localized within the interval of integration. As a rule, an isolated singularity for $\mathbf{Z}(t)'\mathbf{Z}(t)$ corresponds to an isolated singularity in one or more weight functions w_j, and it may be desirable to bypass these by using weight functions sure to be sufficiently smooth. More generally, we may seek weight functions smoother or more regular than those resulting from the pointwise solution.

A strategy for identifying smooth weight functions w_j is to approximate them by using a fixed set of basis functions. Let $\phi_k, k = 1, \ldots, K$ be a set of K such basis functions, and let ϕ denote the K-dimensional vector function $(\phi_1, \ldots, \phi_K)'$. We may use standard basis families such as polynomials, Fourier series, or B-spline functions with a fixed knot sequence, or we may employ a set of basis functions suggested by the application at hand. In any case, we assume that

$$w_j \approx \sum_k c_{jk}\phi_k \tag{14.10}$$

where the mK coefficients c_{jk} define the approximations, and require estimation from the data. Let the (mK)-vector c contain these coefficients, where index k varies inside index j.

We can express or approximate the criterion $\text{SSE}_{\text{PDA}}(L)$ in terms of c as a quadratic form $\hat{F}(c)$ that can be minimized by standard numerical algebraic techniques. We have

$$\hat{F}(c|x) = C + c'\mathbf{R}c + 2c's \tag{14.11}$$

where the constant C does not depend on c, and hence the estimate $\hat{c}$ is given by the solution of the equation $\mathbf{R}c = -s$.

The symmetric matrix **R** in (14.11) is of order mK, and consists of an $m \times m$ array of $K \times K$ submatrices $\mathbf{R}_{jk}$ of the form

$$\mathbf{R}_{jk} = N^{-1} \int \phi(t)\phi(t)' \sum_i D^j x_i(t) D^k x_i(t)\, dt. \tag{14.12}$$

for $0 \le j, k \le m-1$. Similarly, the (mK)-vector s contains m subvectors s_j each of length K, defined as

$$s_j = N^{-1} \int \phi(t) \sum_i D^j x_i(t) D^m x_i(t)\, dt. \tag{14.13}$$

for $j = 0, \ldots, m-1$.

The integrals involved in these expressions often have to be evaluated numerically. For example, it may suffice to use the trapezoidal rule over a fine mesh of equally-spaced values of t.

14.2.3 Regularized PDA

The expansion of w_j in terms of a fixed number of basis functions can be considered a type of regularization process in the sense that we can use the number, choice and smoothness of the basis functions ϕ_k to ensure two potentially desirable features: a smooth or regular variation in w_j, and the closeness of the estimated w_j to some target or hypothesized weight function ω_j.

An alternative type of regularization of w_j, used repeatedly in this book, is to attach penalty terms to the criterion (14.2). One version is

$$\mathsf{PENSSE}_{\mathsf{PDA}}(L) = N^{-1} \sum_i \int Lx_i(t)^2 dt + \sum_{j=0}^{m-1} \lambda_j \int w_j(t)^2 \, dt. \tag{14.14}$$

The scalars λ_j control the roughness of the estimated weight functions. Large values for all λ_j shrink all of the weight functions to 0, with the result that the differential operator converges to $L = D^m$, and the functions that satisfy $Lx = 0$ are the polynomials of degree $m - 1$ or less—an interesting analogy with spline smoothing with operator D^m, where the same functions are regarded as hypersmooth.

If we use the particular roughness penalty of (14.14), then minimizing $\mathsf{PENSSE}_{\mathsf{PDA}}(L)$ is straightforward. One possibility is to proceed pointwise, as in Section 14.2.1, adding the extra term $w(t)'\Lambda w(t)$ to $\mathsf{PSSE}_L(t)$, where $\Lambda = \text{diag}(\lambda_0, \ldots, \lambda_{m-1})$. The resulting estimates of the weight functions are given by

$$\hat{w}(t) = [\mathbf{Z}(t)'\mathbf{Z}(t) + N\Lambda]^{-1}\mathbf{Z}(t)'y(t).$$

Alternatively—whatever the roughness penalty—we can use a basis expansion approach. We add an extra quadratic term to $\hat{F}(c|x)$ in (14.11), corresponding to the expression of the roughness penalty in terms of the coefficients c_{jk}.

14.2.4 Assessing fit

Since the objective of PDA is to minimize the norm $\|L\mathbf{y}\|$ of the forcing function associated with an estimated differential operator, and since the quality of fit can vary over the domain T, it seems appropriate to assess fit in terms of the pointwise error sum of squares $\mathsf{PSSE}_L(t)$ as defined in (14.7). As in linear modelling, the logical baseline against which we should compare PSSE_L is the error sum of squares defined by a theoretical model and its associated weight functions ω_j:

$$\mathsf{PSSE}_0(t) = \sum_i [\sum_{j=0}^{m-1} \omega_j(t)(D^j y_i)(t) + (D^m y_i)(t)]^2. \tag{14.15}$$

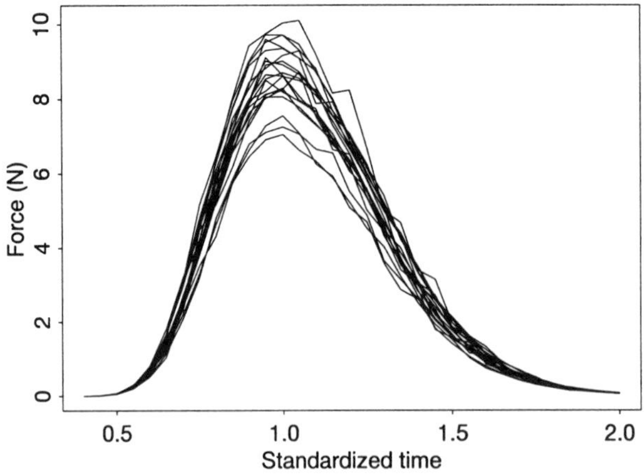

FIGURE 14.7. Twenty recordings of the force exerted by the thumb and forefinger during a brief squeeze of a force meter. The data have been preprocessed to register the functions and remove some shape variability, and the values displayed are for the 33 values $t = 0.4(0.05)2.0$.

In the event that there is no theoretical model at hand, we may use $\omega_j = 0$, so that the comparison is simply with the sum of squares of the $D^m y_i$. From these loss functions, we may examine the pointwise squared multiple correlation function

$$\text{RSQ}(t) = \frac{\text{PSSE}_0(t) - \text{PSSE}_L(t)}{\text{PSSE}_0(t)} \tag{14.16}$$

and the pointwise F-ratio

$$\text{FRATIO}(t) = \frac{(\text{PSSE}_0(t) - \text{PSSE}_L(t))/m}{\text{PSSE}_0(t)/(N-m)}. \tag{14.17}$$

14.3 PDA of the pinch force data

In this section we take up an example in which the estimated linear differential operator is compared with a theoretically defined operator. Further applications of these general ideas are considered in Chapter 15. The data in this example consisted of the 20 records of brief force impulses exerted by the thumb and forefinger in the experiment in motor physiology described in Section 1.4.2. For the purposes of

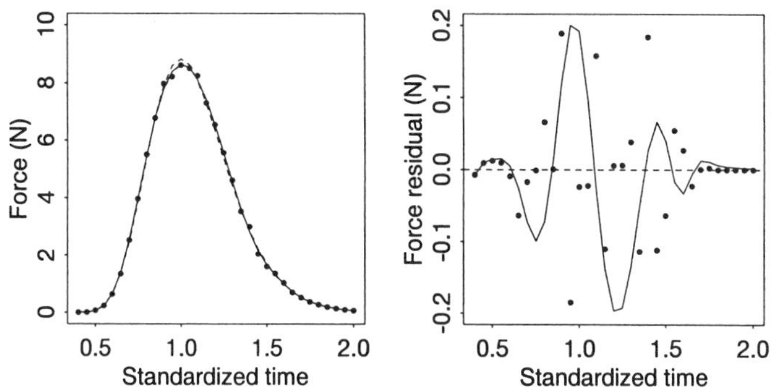

FIGURE 14.8. The left figure contains the data values for the first record (the points), the smoothing spline (solid curve), and the least squares fit by the model (14.18) (dotted curve). The right display shows the residuals arising from fitting the points by a spline function (the points) and the difference between the theoretical model and the spline (solid curve).

this discussion, the force impulses were preprocessed to transform time linearly to a common metric and to remove some simple shape variation. The resulting curves are displayed in Figure 14.7. Details of the preprocessing stages can be found in Ramsay, Wang and Flanagan (1995).

There are some theoretical considerations which suggest that the model

$$y_i(t) = C_i \exp[-\log^2 t/(2\sigma^2)] \tag{14.18}$$

offers a good account of any specific force function. In this application, the data were preprocessed to conform to a fixed shape parameter σ^2 of 0.05. Functions of the form (14.18) are annihilated by the differential operator $L_0 = [(t\sigma)^{-1} \log t]I + D$. A goal of this analysis is to compare this theoretical operator with the first order differential operator $L = w_0 I + D$ estimated from the data, or to compare the theoretical weight function $\omega_0(t) = (t\sigma)^{-1} \log t$ with its empirical counterpart w_0.

We spline smoothed the records, penalizing by the integral of the square of the third derivative to get a smooth first derivative estimate. It is clear from Figure 14.7 that the size of error variation is not constant over time. Accordingly, we estimated the residuals in a first smoothing step, and smoothed the logs of their standard deviations to estimate the variation of the typical residual size over time. Then we took the

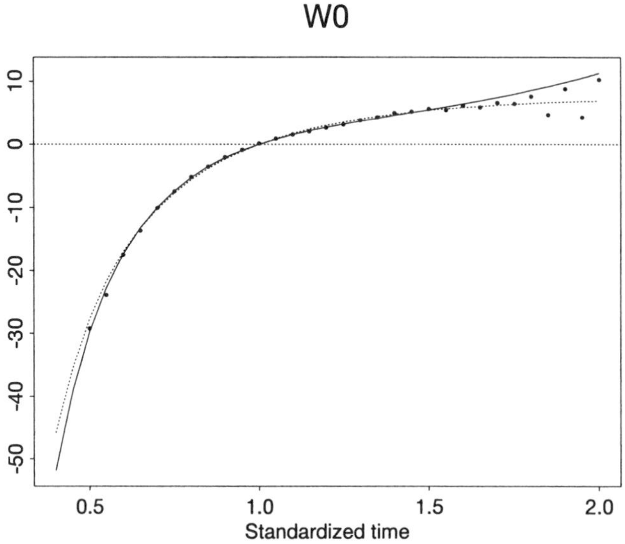

FIGURE 14.9. The weight function estimated by the basis expansion method for the pinch force data is indicated by the solid line, the theoretical function by the dotted line, and the pointwise estimates by the dots.

weights σ_j^2 in the weighted spline smoothing criterion

$$\text{PENSSE}_\lambda(x|y) = \sum_j \{y_j - x(t_j)\}^2/\sigma_j^2 + \lambda \|D^3 x\|^2 \tag{14.19}$$

to be the squares of the exponential-transformed smooth values. Finally, we re-smoothed the data to get the spline smoothing curves and their derivatives.

Figure 14.8 displays the discrete data points, the smoothing function, and also the theoretical function (14.18) fit by least squares for a single record. The theoretical function fits very well, but in the right panel we see that the discrepancy between the theoretical model and the smoothing spline fit is itself smooth and of the same order as the largest deviations of the points from this flexible spline fit. Although this discordance between the model and the spline is less than 2% of the size of the force itself, we are entitled to wonder if this theoretical model can be improved.

We applied both the pointwise and basis expansion procedures for estimating w_0 to the smooth functions and their derivatives. The basis used for the basis expansion procedure was

$$\phi(t) = (t^{-1}\log t, 1, t-1, (t-1)^2)',$$

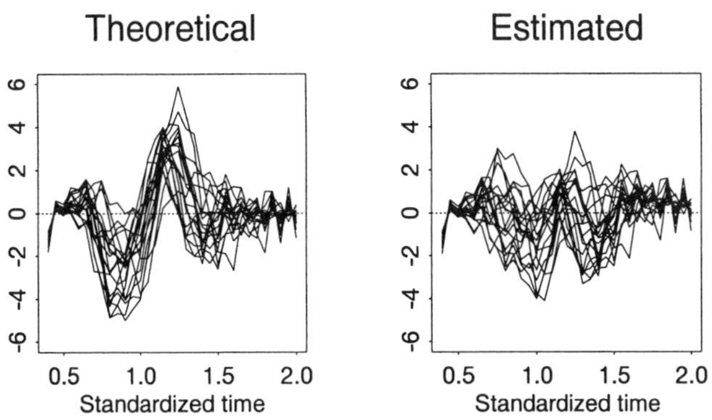

FIGURE 14.10. The left panel displays the forcing or impulse functions Ly_i produced by the theoretical operator, and the right panel shows the corresponding empirical operator functions.

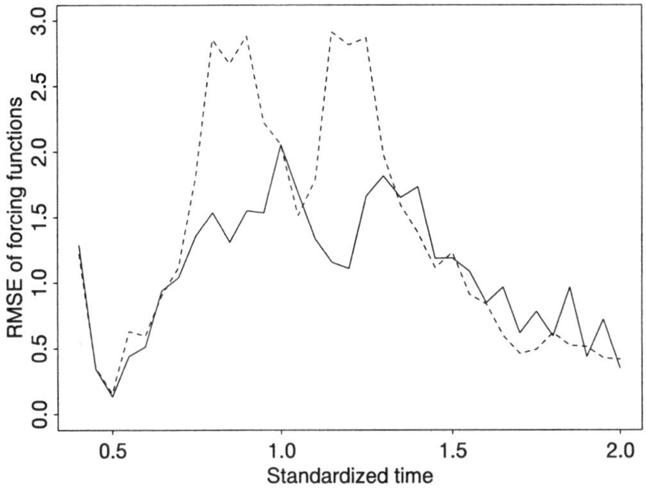

FIGURE 14.11. The solid line indicates the square root of the mean squared forcing function for the estimated operator, and the dotted line the same quantity for the theoretical operator.

chosen after some experimentation; the first basis function was suggested by the theoretical model, and the remaining polynomial terms served to extend this model as required. Figure 14.9 shows the theoretical, the pointwise, and the global estimates of the weight functions. These are admittedly close to one another, at least in the central ranges of adjusted time, but again we observe some slight but consistent differences between the theoretical and empirical weight functions.

However, the forcing functions Ly_i, displayed in Figure 14.10, show a systematic trend for the theoretical operator, whereas the empirical forcing functions exhibit much less pattern. Figure 14.11 displays the root-mean-squares of the two sets of forcing functions, and this permits us to see more clearly that the estimated operator is superior in the periods just before and after the peak force, where it produces a forcing function about half the size of its theoretical counterpart. It seems appropriate to conclude that the estimated operator has produced an important improvement in fit on either side of the time of maximum force. Ramsay, Wang and Flanagan (1995) conjecture that the discrepancy between the two forcing functions is due to drag or viscosity in the thumb–forefinger joint.

15

More general roughness penalties

15.1 Introduction

15.1.1 More general roughness penalties and L-splines

A theme central to this book has been the use of roughness penalties to incorporate smoothing, whether in the context of using discrete data to define a smooth function in Chapter 3, functional principal components analysis in Chapter 7, or imposing regularity on estimated regression functions in Chapter 10 or on canonical variate weight functions in Chapter 12.

At the same time, the previous two chapters have dealt with the mathematical properties of linear differential operators L and with techniques for estimating them from data. Principal differential analysis provides a method of estimating low-dimensional functional variation in a sense analogous to principal components analysis, but by estimating an m-th order differential operator L rather than a projection.

Moreover, we have seen that by coupling L with a suitable set of constraints on the m linearly independent functions u_j satisfying $Lu_j = 0$, we can *partition* the space of smooth functions into two parts. This is achieved by defining a constraint operator B such that $Bu_j \neq 0$, and that the only function satisfying $Bx = Lx = 0$ is $x = 0$. Then any function x having m derivatives can be expressed uniquely as

$$x = u + e \text{ where } Lu = 0 \text{ and } Be = 0. \tag{15.1}$$

We might call this the *partitioning principle.*

It is time to put these two powerful ideas together, to see what practical value there is in using the partitioning principle to define a roughness penalty. We want to go beyond the standard practice of defining roughness in terms of $L = D^2$, and even beyond the slightly more general $L = D^m$, to consider what the advantages of using an arbitrary operator L might be, perhaps in conjunction with some constraints captured in the companion operator B. Specifically, when the goal is smoothing the data, we propose using the criterion

$$\text{PENSSE}(x) = \sum_j^n [y_j - x(t_j)]^2 + \lambda \times \text{PEN}_L(x) \tag{15.2}$$

where

$$\text{PEN}_L(x) = \int (Lx)^2(t)\, dt.$$

We begin with some examples.

15.1.2 The lip movement data

Consider the lip movement data introduced in Chapter 14.1.2 and plotted in Figure 14.1. We are interested in how these trajectories, all based on observations of a speaker saying "bob", vary from one replication to another. But in the experiment, the syllable was embedded in the phrase, "Say bob again," and it is clear that the lower lip enters and leaves the period during which the syllable is being formed at different heights. This is nuisance variation that we would be happy to eliminate.

Moreover, there was particular interest in the acceleration or second derivative of the lip, suggesting that we should penalize the fourth derivative by spline smoothing with $L = D^4$. Any cubic polynomial trend in the records is ignored if we do that. Now we want to define the *shape* component u and *endpoint* component e of each record x so that the behaviour of the record at the beginning and end of the interval of observation (normalized to be [0,1]) has minimal impact on the interior and more interesting portion of the curve. One way of achieving this objective is to require that shape components satisfy the constraints $e(0) = De(0) = 0$ and $e(1) = De(1) = 0$. This means that the chosen constraint operator is B_E, defined as

$$B_E x = [x(0), Dx(0), x(1), Dx(1)]', \tag{15.3}$$

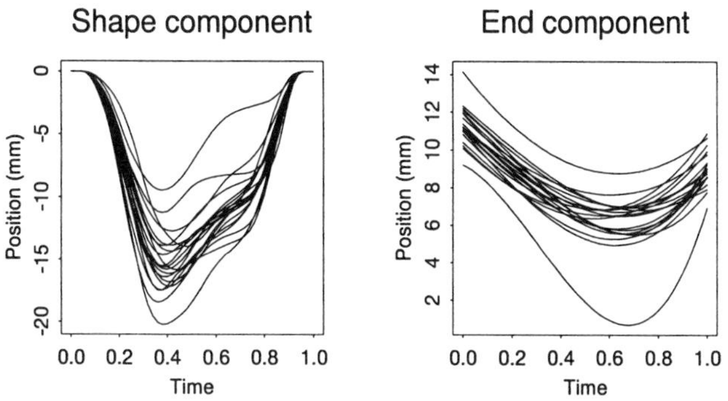

FIGURE 15.1. The right panel displays the 20 cubic polynomials u that match the lip position and derivative values at 0 and 1 for the smoothed versions of the curves in Figure 14.1. The left panel shows the shape components e that have zero endpoint positions and derivatives.

and the shape component e satisfies $B_E e = 0$.

We now have our two linear operators $L = D^4$ and $B = B_E$ in hand, and they are complementary in the sense that $\ker B \cap \ker L = 0$. We can therefore unambiguously split any lip position record x into $x = u + e$, where $Be = 0$, and u, a cubic polynomial because $Lu = D^4 u = 0$, picks up the endpoint variation by fitting the record's function and derivative values at both 0 and 1. Figure 15.1 displays the endpoint and shape components for all 20 records.

15.1.3 The weather data

In the introduction we noted that a rather large part of the mean daily or monthly temperature curve for any weather station can be captured by the simple function

$$T(t) = c_1 + c_2 \sin(\pi t/6) + c_3 \cos(\pi t/6) \tag{15.4}$$

and the same may be said for the log precipitation profiles. Functions of this form can be annihilated by the operator

$$L = (\pi/6)^2 D + D^3.$$

We could propose smoothing data using the criterion (15.2), where

$$\text{PEN}_L(x) = \int_0^{12} (Lx)^2(t)\, dt = \int_0^{12} \{(\pi/6)^2 Dx(t) + D^3 x(t)\}^2\, dt$$

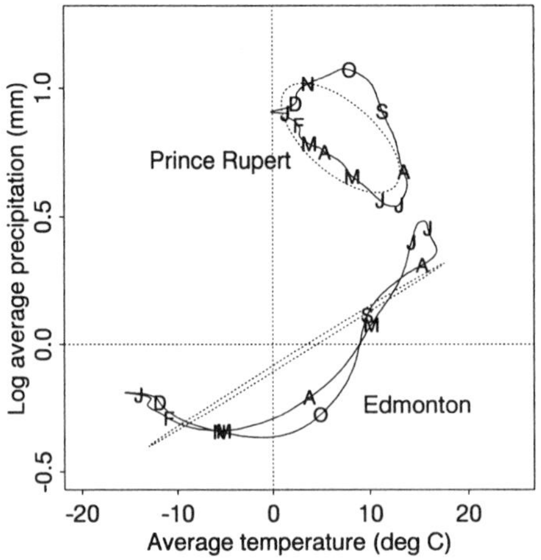

FIGURE 15.2. The solid cycles are the smoothed daily temperature and log precipitation data, plotted against each other, for two Canadian weather stations. The dotted curves are the estimated cycles based on strictly sinusoidal variation, taking the first three terms of the Fourier expansion of each observed temperature and log precipitation curve. Letters indicate the middle of each month.

while paying attention to the periodic character of the data. What would we gain from this? For one thing, as we have already noted in Section 13.3.2, this procedure is likely to have considerable advantages in estimating curves x from raw data.

At the same time, the function LTemp in this example is interesting in itself, and Ramsay and Dalzell (1991) refer to it as the *harmonic acceleration* of temperature. By functional principal components and linear regression analyses, they show that LTemp and the harmonic acceleration of log precipitation contain a great deal of information about the peculiarities of weather at any station. To identify the component e uniquely, though, we must choose a matching constraint operator B, and for this application they chose

$$Bx = [\int x(t)dt, \int x(t)\sin(\pi t/6)dt, \int x(t)\cos(\pi t/6)dt]'$$

corresponding to the first three Fourier coefficients of the observed curves. The three functions u_i that span $\ker B$ are then 1, $\sin(\pi t/6)$ and $\cos(\pi t/6)$. Given any curve x, the partition (15.1) is achieved by setting the component u to be the first three terms in the Fourier expansion of x.

For two weather stations, the solid curves in Figure 15.2 show plots of smoothed daily temperature against smoothed daily log precipitation through the year. When plotted against one another, the shifted sinusoidal components $u(t)$ for temperature and log precipitation for each station become ellipses, shown as dotted curves.

15.2 The optimal basis for spline smoothing

In Section 3.2 we reviewed the classic technique of representing functions by fitting a basis function expansion to the data. We took pains to point out that not all bases are equal; a good basis consists of functions which mimic the general features that we know apply to the data, such as periodicity, asymptotic linearity, and so on. When we get these features right, we can expect to do a good job with a smaller number of basis functions.

We also pointed out that when the number n of data points is large, computing an expansion in $O(n)$ operations is critical, and to achieve this, the basis functions should be nonzero only locally, or have compact support. The B-spline basis is especially attractive from this perspective.

In Section 4.1, we extended the basis function expansion concept to employ a partitioned basis (ϕ, ψ) along with a penalty on the size of the component expanded in terms of the basis functions ψ. But two properties, relevance to the data and convenience of computation, remain essential.

We now bring these elements together. We use the partitioning principle to define a set of basis functions that are optimal with respect to smoothing. We then provide a recipe for an $O(n)$ smoothing algorithm, and also show how the basis can be chosen to be of compact support form to give the appropriate analogue of B-splines.

We begin with a theorem stating that the optimal basis for spline smoothing in the context of operators (B, L) is defined by the reproducing kernel k_2 defined in Chapter 13.

Optimal Basis Theorem: *For any $\lambda > 0$, the function x minimizing the spline smoothing criterion (15.2) defined by a linear differential operator L of order m has the expansion*

$$x(t) = \sum_{j=1}^{m} d_j u_j(t) + \sum_{i=1}^{n} c_i k_2(t_i, t). \tag{15.5}$$

Equation (15.5) can be put a bit more compactly. As before, let $u = (u_1, \ldots, u_m)'$. Define another vector function

$$\tilde{k}(t) = \{k_2(t_1, t), k_2(t_2, t), \ldots, k_2(t_n, t)\}'.$$

Then the optimal basis theorem says that the function x has to be of the form $x = d'u + c'\tilde{k}$, where d is a vector of m coefficients d_j and c is the corresponding vector of n coefficients c_i in (15.5). We give a proof of the optimal basis theorem, but as usual any reader prepared to take this on trust should simply skip to the next section.

Proof: Suppose x^* is any function having square-integrable derivatives up to order m. The strategy for the proof is to construct a function $\tilde{x}$ of the form (15.5) such that $\text{PENSSE}(\tilde{x}) \le \text{PENSSE}(x^*)$ with equality only if $\tilde{x} = x^*$. It follows at once that we need never look beyond functions of the form (15.5) to minimize the spline smoothing criterion PENSSE.

First of all, write $x^* = u^* + e^*$ where $u^* \in \ker L$ and $e^* \in \ker B$. Let $\mathcal{K}$ be the subspace of $\ker B$ spanned by the n functions $k_2(t_i, \cdot)$, and let $\tilde{e}$ be the projection of e^* onto $\mathcal{K}$ in the L-inner product, as defined in Section 13.7. This means that $e^* = \tilde{e} + e^\perp$, where $\tilde{e} = c'\tilde{k}$ for some vector c, and the residual $e^\perp$ in $\ker B$ satisfies the orthogonality condition

$$\langle e, e^\perp \rangle_L = \int (Le)(Le^\perp) = 0 \text{ for all } e \text{ in } \mathcal{K}. \tag{15.6}$$

We now define our function $\tilde{x} = u^* + \tilde{e}$, so that $\tilde{x}$ is necessarily of the required form (15.5), and $x^* - \tilde{x}$ is equal to the residual $e^\perp$.

To show that $\text{PENSSE}(\tilde{x}) \le \text{PENSSE}(x^*)$, first note that for each i, by the defining property of the reproducing kernel,

$$x^*(t_i) - \tilde{x}(t_i) = e^\perp(t_i) = \langle k_2(t_i, \cdot), e^\perp \rangle_L = 0$$

by property (15.6), since $k_2(t_i, \cdot)$ is a member of $\mathcal{K}$ and so is L-orthogonal to $e^\perp$.

Therefore x^* and $\tilde{x}$ agree at the arguments t_i, and so

$$\text{PENSSE}(x^*) - \text{PENSSE}(\tilde{x}) = \lambda\{\text{PEN}_L(x^*) - \text{PEN}_L(\tilde{x})\};$$

the residual sum of squares of the y_i is the same about each of the two functions x^* and $\tilde{x}$. Since $Lx^* = Le^*$ and $L\tilde{x} = L\tilde{e}$, we have

$$\begin{aligned} \text{PEN}_L(x^*) - \text{PEN}_L(\tilde{x}) &= \text{PEN}_L(e^*) - \text{PEN}_L(\tilde{e}) \\ &= \langle \tilde{e} + e^\perp, \tilde{e} + e^\perp \rangle_L - \langle \tilde{e}, \tilde{e} \rangle_L = \langle e^\perp, e^\perp \rangle_L + 2\langle \tilde{e}, e^\perp \rangle_L = \langle e^\perp, e^\perp \rangle_L \end{aligned}$$

since $\tilde{e}$ is in $\mathcal{K}$ and, therefore, is L-orthogonal to $e^\perp$. So $\text{PEN}_L(e^*) \ge \text{PEN}_L(\tilde{e})$, and consequently $\text{PENSSE}(x^*) \ge \text{PENSSE}(\tilde{x})$. Equality holds only if $e^\perp \in \ker L$; since we already know that $e^\perp \in \ker B$, this implies that $e^\perp = 0$ and that $x^* = \tilde{x}$, completing the proof of the theorem.

15.3 An $O(n)$ algorithm for L-spline smoothing

15.3.1 The need for a good algorithm

In principle, the optimal basis theorem should tell us exactly how to proceed. Since we know that the required function has the form $x = d'u + c'\tilde{k}$, we need only express $\mathsf{PENSSE}(x)$ in terms of c and d and minimize to find the best values of c and d. How would this work out?

Let $\mathbf{K}$ be the matrix with values $k_2(t_i, t_j)$. From equation (13.39) it follows that

$$\mathsf{PEN}_L(x) = \langle c'\tilde{k}, c'\tilde{k}\rangle_L = c'\mathbf{K}c.$$

The vector of values $x(t_i)$ is $\mathbf{U}d + \mathbf{K}c$, where $\mathbf{U}$ is the matrix with values $u_j(t_i)$. Hence, at least in principle, we can find x by minimizing the quadratic form

$$\mathsf{PENSSE}(x) = (y - \mathbf{U}d - \mathbf{K}c)'(y - \mathbf{U}d - \mathbf{K}c) + \lambda c'\mathbf{K}c \tag{15.7}$$

to find the vectors c and d.

Unfortunately, in practice the matrix $\mathbf{K}$ is usually extremely badly conditioned, that is to say, the ratio of its largest eigenvalue to its smallest explodes. A practical consequence of this is that the computations required to minimize the quadratic form (15.7) are likely to be unstable or impossible.

Furthermore, in smoothing long sequences of observations, it is crucial to devise a smoothing procedure requiring a number of arithmetic operations that does not grow too quickly as the length of the sequence increases. For example, the handwriting data has $n = 1401$, and so an algorithm that was $O(n^2)$ would be impracticable and an $O(n^3)$ algorithm virtually impossible with current computing power. By adopting a somewhat different approach, we can set out an algorithm that requires only $O(n)$ operations, and furthermore avoids the numerical problems inherent in the direct minimization of (15.7).

The algorithm we use is based on the theoretical paper of Anselone and Laurent (1967), but is also known as the Reinsch algorithm because of the application to the cubic polynomial smoothing case ($L = D^2$) by Reinsch (1967, 1970). It was subsequently extended by Hutchison and de Hoog (1985). We do not attempt a full exposition of the rationale for this algorithm here, but Heckman and Ramsay (1996) and Ramsay, Heckman and Silverman (1997) can be consulted for details.

The algorithm requires evaluations of two types of function that we have already encountered:

1. $u_j, j = 1, \ldots, m$: a set of m linearly independent functions satisfying $Lu_j = 0$, that is, spanning $\ker L$. As before, we refer to these collectively as the vector-valued function u.

2. k_2: the reproducing kernel function defined in Chapter 13 for the subspace of functions e satisfying $B_I e = 0$, where B_I is the initial value constraint operator.

The functions u and k_2 are the user-supplied components of the algorithm and are defined by the particular choice of operator L used in the smoothing application.

The algorithm splits into three phases:

1. an initial setup phase that does not depend on the smoothing parameter λ,

2. a smoothing phase in which we smooth the data, and

3. a summary phase in which we compute performance measures for the smooth.

This division of the task is of practical importance because we may want to try smoothing with many values of λ, and do not want needlessly to repeat either the initial setup phase or the final descriptive phase.

15.3.2 Setting up the smoothing procedure

In the initial phase, we define two symmetric $(n - m) \times (n - m)$ band-structured matrices $\mathbf{H}$ and $\mathbf{C}'\mathbf{C}$ where m is the order of the operator L. Both matrices are band-structured with band width at most $2m + 1$, which means that all entries more than m positions away from the main diagonal are zero. Because of symmetry, these band-structured matrices require only $(n - m)(m + 1)$ storage locations.

We start by explaining how to construct the matrix $\mathbf{C}$. For each $i = 1, \ldots, n - m$, define the $(m+1) \times m$ matrix $\mathbf{U}^{(i)}$ to have (l, j) element $u_j(t_{i+l})$, for $l = 0, \ldots, m$. Thus $\mathbf{U}^{(i)}$ is the submatrix of $\mathbf{U}$ consisting only of rows $i, i+1, \ldots, i+l$. Find the QR decomposition (as discussed in Section A.1.3)

$$\mathbf{U}^{(i)} = \mathbf{Q}^{(i)}\mathbf{R}^{(i)}$$

where $\mathbf{Q}^{(i)}$ is square, of order $m + 1$, and orthonormal, and $\mathbf{R}^{(i)}$ is $(m+1) \times m$ and upper triangular. Let the vector $c^{(i)}$ be the last column of $\mathbf{Q}^{(i)}$; this vector is orthogonal to all the columns of $\mathbf{U}^{(i)}$. In fact any vector having this property will do, and in special cases the vector can be found by some other method. For polynomial spline smoothing, for instance, coefficients defining divided differences are used.

Now define the $n \times (n-m)$ matrix $\mathbf{C}$ so that its ith column has the $m+1$ values $c^{(i)}$ starting in row i; elsewhere the matrix contains zeroes. In practice, the argument sequence $t_1, \ldots, t_n$ is often equally spaced, and in this case it frequently happens that all the coefficient vectors $c^{(i)}$ are the same, and hence need be computed only once. The band structure of $\mathbf{C}$ immediately implies that $\mathbf{C}'\mathbf{C}$ has the required band structure, and can be found in $O(n)$ operations for fixed m.

The other setup-phase matrix $\mathbf{H}$ is the $(n-m) \times (n-m)$ symmetric matrix

$$\mathbf{H} = \mathbf{C}'\mathbf{K}\mathbf{C}, \tag{15.8}$$

where $\mathbf{K}$ is the matrix of values $k_2(t_i, t_j)$. It turns out that $\mathbf{H}$ is also band-structured with band width $2m-1$. This is a consequence of the expression (13.41) for the reproducing kernel, which yields the following two-part expression:

$$k_2(t_i, t) = \begin{cases} u(t_i)'\mathbf{F}(t)u(t) & \text{for } t_i \geq t \\ u(t_i)'\mathbf{F}(t_i)u(t) & \text{for } t_i \leq t, \end{cases} \tag{15.9}$$

for a certain matrix function $\mathbf{F}(t)$. This implies that

$$\mathbf{K}_{ij} = \{\mathbf{U}\mathbf{F}(t_j)u(t_j)\}_i \text{ for } i \geq j. \tag{15.10}$$

Suppose $k \geq j$. Because $\mathbf{C}_{ik}$ is zero for $i < k$,

$$(\mathbf{C}'\mathbf{K})_{kj} = \sum_{i=k}^{n} \mathbf{C}_{ik}\mathbf{K}_{ij} = \sum_{i=k}^{n} \mathbf{C}_{ik}\{\mathbf{U}\mathbf{F}(t_j)u(t_j)\}_i,$$

substituting (15.10); notice that $i \geq j$ for all i within the range of summation $k \leq i \leq n$. It follows that for $k \geq j$ we have

$$(\mathbf{C}'\mathbf{K})_{kj} = \{\mathbf{C}'\mathbf{U}\mathbf{F}(t_j)u(t_j)\}_i = 0.$$

So $\mathbf{C}'\mathbf{K}$ is strictly upper-triangular. Because of the band structure of $\mathbf{C}$, the matrix $\mathbf{H} = (\mathbf{C}'\mathbf{K})\mathbf{C}$ has zero entries for positions m or more below the main diagonal, and by symmetry $\mathbf{H}$ has the stated band structure.

15.3.3 *The smoothing phase*

The actual smoothing consists of two steps:

1. Compute the $(n-m)$-vector z that solves

$$(\mathbf{H} + \lambda\mathbf{C}'\mathbf{C})z = \mathbf{C}'y, \tag{15.11}$$

where the vector y contains the values to be smoothed.

2. Compute the vector of n values $\hat{y}_i = x(t_i)$ of the smoothing function x at the n argument values using

$$\hat{y} = y - \lambda \mathbf{C} z. \tag{15.12}$$

Because of the band structure of $(\mathbf{H} + \lambda \mathbf{C}'\mathbf{C})$ and of $\mathbf{C}$, both of these steps can be computed in $O(n)$ operators, and references on efficient matrix computation such as Golub and van Loan (1989) can be consulted for details.

15.3.4 *The performance assessment phase*

The vector of smoothed values $\hat{y}$ and the original values y that were smoothed are related as follows:

$$\begin{aligned} \hat{y} &= y - \lambda \mathbf{C}(\mathbf{H} + \lambda \mathbf{C}'\mathbf{C})^{-1}\mathbf{C}' y \\ &= \{\mathbf{I} - \lambda \mathbf{C}(\mathbf{H} + \lambda \mathbf{C}'\mathbf{C})^{-1}\mathbf{C}'\} y. \end{aligned} \tag{15.13}$$

The matrix $\mathbf{S}$ defined by

$$\mathbf{S} = \mathbf{I} - \lambda \mathbf{C}(\mathbf{H} + \lambda \mathbf{C}'\mathbf{C})^{-1}\mathbf{C}' \tag{15.14}$$

is often called the *hat matrix*, and in effect defines a linear transformation that maps the unsmoothed data into its smooth image by

$$\hat{y} = \mathbf{S} y.$$

Various measures of performance depend on the diagonal values in $\mathbf{S}$. Of these, the most popular are currently

$$\mathsf{GCV} = \mathsf{SSE}/(1 - n^{-1} \operatorname{trace} \mathbf{S})^2 \tag{15.15}$$

where

$$\mathsf{SSE} = \sum_{i=1}^{n} [y_i - x(t_i)]^2 = \|y - \hat{y}\|^2$$

and

$$\mathsf{CV} = \sum_{i=1}^{n} [\{y_i - x(t_i)\}/(1 - s_{ii})]^2 \tag{15.16}$$

where s_{ii} is the ith diagonal entry of $\mathbf{S}$. Using methods developed by Hutchison and de Hoog (1985), we can compute both measures GCV and CV in $O(n)$ operations because of the band-structured nature of the matrices defining $\mathbf{S}$, One of the main applications of these two criteria, both of which are types of discounted error sums of squares, is as a guide for choosing the value of the smoothing parameter λ. It is relatively standard practice to look for the value that minimizes one of these two criteria, just as various variable selection procedures attempt to minimize discounted error sums of squares in standard regression analysis. Interestingly, the GCV measure was originally introduced by Craven and Wahba (1979) as an approximation to the CV criterion that could be computed in $O(n)$ operations. Now CV is also available in n operations, but GCV still tends to be preferred in practice for other reasons. For example, various simulation studies have indicated that GCV tends to give a better choice of the smoothing parameter λ, possibly because GCV makes use of smoothing itself by replacing the variable values $1 - s_{ii}$ with the average $1 - n^{-1}\text{trace}\,\mathbf{S}$.

Also of great value is a measure of the effective number of degrees of freedom of the smoothing operation. Two measures are

$$\mathsf{DF}_1 = \text{trace}\,\mathbf{S} \text{ and } \mathsf{DF}_2 = \text{trace}\,\mathbf{S}'\mathbf{S} = \text{trace}\,\mathbf{S}^2. \tag{15.17}$$

These dimensionality measures were introduced and discussed by Buja et al. (1989). It can be shown that, in the limit as $\lambda \to \infty$, both measures become simply m, and similarly, as $\lambda \to 0$, both measures converge to n. In between, they give slightly different impressions of how much of the variation in the original unsmoothed data remains in the smoothed version.

15.3.5 Other $O(n)$ algorithms

There is an intimate connection between the theory of splines and the theory of stochastic differential equations (Wahba, 1978, 1990, Weinert, Byrd and Sidhu, 1980). This leads to the possibility of using the Kalman filter, a technique widely used in engineering and other fields for extracting an estimate of a signal from noisy data, to compute a smoothing spline. Ansley, Kohn and Wong (1993), using a Kalman filtering algorithm described in Ansley and Kohn (1989), give some examples of computing an L-spline in $O(n)$ operations. However, except for fairly simple cases, this algorithm appears to be difficult to implement, and its description involves substantial mathematical detail. Nevertheless, we feel that it is important to call the reader's attention to this stimulating literature on smoothing by state-space methods.

15.4 A compact support basis for L-splines

In this section our concern is the construction of compact support basis functions from the reproducing kernel basis functions $k_2(t_i, \cdot)$. A basis made up of such functions may be useful for techniques such as the regularized principal components analysis described in Section 7.4.1, and has many numerical advantages, analogous to those of B-splines.

For any fixed $i = 1, \ldots, n - 2m$, consider the following sequence of $2m + 1$ basis functions based on the reproducing kernel:

$$k_2(t_{i+\ell}, \cdot), \ell = 0, \ldots, 2m.$$

Let $b_\ell^{(i)}, \ell = 0, \ldots, 2m$ be a corresponding sequence of weights defining a new basis function

$$\psi_i = \sum_{\ell=0}^{2m} b_\ell^{(i)} k_2(t_{i+\ell}, \cdot). \tag{15.18}$$

The properties we seek are

$$\psi_i(t) = 0 \text{ for } t \le t_i, \text{ and } \psi_i(t) = 0 \text{ for } t \ge t_{i+2m}.$$

But from the first line of (15.9), the first of these is achieved if

$$\sum_{\ell=0}^{2m} b_\ell^{(i)} u_{j_1}(t_{i+\ell}) = 0, \; j_1 = 1, \ldots, m \tag{15.19}$$

and, at the same time, the second line of (15.9) indicates that the second property is satisfied if

$$\sum_{\ell=0}^{2m} b_\ell^{(i)} [\sum_{j_1=1}^{m} u_{j_1}(t_{i+\ell}) f_{j_1,j_2}(t_{i+\ell})] = 0, \; j_2 = 1, \ldots, m \tag{15.20}$$

where $f_{j_1,j_2}(t_{i+\ell})$ is entry (j_1, j_2) of $\mathbf{F}(t_{i+\ell})$.

Now these are two sets of m linear constraints on the $2m + 1$ coefficients $b_\ell^{(i)}$, and we know that, in general, we can always find a coefficient vector $b^{(i)}$ satisfying them. The reason that there are only $2m$ constraints for $2m+1$ coefficients is that the linear constraints can only define the vector $b^{(i)}$ up to a change of scale.

Let the $(2m + 1) \times 2m$ matrix $\mathbf{V}^{(i)}$ have in its first m columns the values $u_{j_1}(t_{i+\ell}), j_1 = 1, \ldots, m$, and in its second set of m columns the values $\sum_{j_1=1}^{m} u_{j_1}(t_{i+\ell}) f_{j_1,j_2}(t_{i+\ell}), j_2 = 1, \ldots, m$. Then the constraints (15.19) and (15.20) can be written in the matrix form

$$(b^{(i)})' \mathbf{V}^{(i)} = 0.$$

As in the calculation of the vectors $c^{(i)}$ in Section 15.3.2, the required vector $b^{(i)}$ is simply the last column of the **Q** matrix in the QR decomposition of $\mathbf{V}^{(i)}$.

If the argument values are unequally spaced, this calculation of the coefficient vectors $b^{(i)}$ must be carried out for each value of i from 1 to $n - 2m$. However, in the frequently encountered case where the t_i values are equally spaced, only one coefficient calculation is required, and the resulting set of coefficients b serves for all $n - 2m$ compact support splines ψ_i.

Observant readers may note that we have lost $2m$ basis functions by this approach. We may deal with this difficulty in various ways. One approach is to say that a little bit of fitting power has been lost, but that, if n is large, this may have little impact on the smoothing function, and what little impact it has is at the ends of the interval. Alternatively, however, we can use a technique employed in defining polynomial B-splines, and add m additional argument values at each end of the interval. For computational convenience in the equally-spaced argument case, we can simply make these a continuation of the sequence in both directions. This augments the basis to retain the full fitting power of the original reproducing kernel basis.

15.5 Some case studies

15.5.1 *The gross domestic product data*

The gross domestic product data introduced in Chapter 13 share with many economic indicators the overall tendency for exponential growth. If we wish to smooth the deseasonalized GDP record of the U.S. displayed in Figure 13.1, the operator $L = -\gamma D + D^2$ annihilates $u_1(t) = 1$ and $u_2(t) = e^{\gamma t}$, so these are obvious choices for the functions spanning $\ker L$. A reasonable choice for the matching constraint operator is simply B_I, such that $B_I u = \{u(0), Du(0)\}'$, implying that the coefficients of u_1 and u_2 are specified by the initial value and slope of the smoothed record.

In this case, we could estimate the parameter γ by estimating the slope of the relationship between log GDP and time by ordinary regression analysis. Another possibility is to fit all or part of the data by nonlinear least squares regression using the two functions u_1 and u_2. To do this, we minimize the error sum of squares with respect to the coefficients c_1 and c_2 of $c_1u_1 + c_2u_2$ and with respect to γ which, of course, determines u_2. Since for any fixed γ value, the minimizing

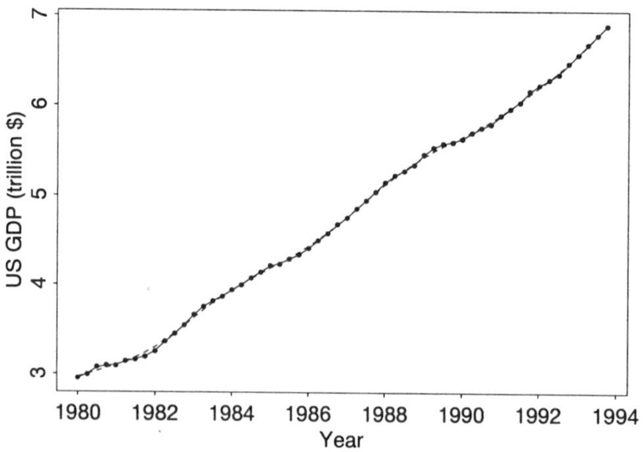

FIGURE 15.3. The line virtually interpolating the data indicates the spline smooth of the U.S. GDP data using $L = -0.054D + D^2$ and the minimum GCV value for smoothing parameter λ. The dashed line, which is almost indistinguishable but does not quite interpolate the data so closely, indicates the L-spline fit corresponding to $\mathrm{DF}_1 = 10$.

values of the coefficients can be computed directly by linear least squares, it makes sense to use a one-dimensional minimization routine such as Brent's method (Press et al., 1992) to find the optimal γ value; for each new value of γ within the iterative method, a linear regression is required to get the associated values of c_1 and c_2. The resulting least squares estimate of γ for the U.S. data, based on the values from 1980 to 1989 when the growth was more exponential, is 0.054.

Using this value of γ, we used the method of Section 15.3 to find the L-smoothing spline shown in Figure 15.3. We minimized the GCV criterion to obtain the value $\lambda = 0.053$. The DF_1 measure of equivalent degrees of freedom was 39.6, so we purchased the excellent fit of the spline at the price of a rather large number of degrees of freedom.

By comparison, the cubic smoothing spline that minimizes GCV produces almost identical results in terms of GCV and DF_1 values. Perhaps this is not surprising since the curve is only slightly more exponential than linear. But the results are rather different when we smooth with the fixed value of $\mathrm{DF}_1 = 10$, corresponding to $\lambda = 22.9$. The L-spline fit using this more economical model is just barely visible in Figure 15.3, and GCV = 0.00068. The cubic polynomial spline with

$\mathsf{DF}_1 = 10$ yields $\mathsf{GCV} = 0.00084$, and its poorer fit reflects the fact that some of its precious degrees of freedom were used up in fitting the mild exponential trend.

15.5.2 The melanoma data

These data, displayed in Figure 13.2, represent a more complex relationship, with a cyclic effect superimposed on a linear development. The interesting operator is

$$L = \omega^2 D^2 + D^4 \tag{15.21}$$

for some appropriate constant ω, since this would annihilate the four functions

$$u(t) = (1, t, \sin \omega t, \cos \omega t)'.$$

Using the techniques of Chapter 13, for $s \leq t$ the reproducing kernel is given by

$$\begin{aligned} k_2(s,t) &= \omega^{-7}[(\omega s)^2(\omega t/2 - \omega s/6) - \omega t + \omega s + \omega t \cos \omega s \\ &\quad + \omega s \cos \omega t - \sin \omega s - \sin \omega t + \sin(\omega t - \omega s) \\ &\quad - (\sin \omega s \cos \omega t)/2 + s \cos(\omega t - \omega s)/2]. \end{aligned} \tag{15.22}$$

We estimated the parameter ω to be 0.650 by the nonlinear least squares approach. This corresponds to a period of 9.66 years, roughly the period of the sunspot cycle affecting solar radiation and consequently melanoma. When we smooth the data with the spline defined by the operator (15.21) and select λ so as to minimize GCV, it turns out that λ becomes arbitrarily large, corresponding to a parametric fit using only the basis functions u, consuming four degrees of freedom, and yielding $\mathsf{GCV} = 0.076$. The polynomial smoothing spline with order $m = 4$, displayed in Figure 13.2, is a minimum-GCV estimate corresponding to $\mathsf{DF}_1 = 12.0$ and $\mathsf{GCV} = 0.095$. Thus, polynomial spline smoothing required three times the degrees of freedom to produce a fit that was still worse in GCV terms than the L-spline smooth. Of the two methods of order four, the operator (15.21) is much to be preferred to $L = D^4$.

15.5.3 GDP data with seasonal effects

In the data provided by the U.S. and most other countries, the within-year variation in GDP that is a normal aspect of most economies was removed. But the data for Sweden, displayed in Figure 15.4, retains this

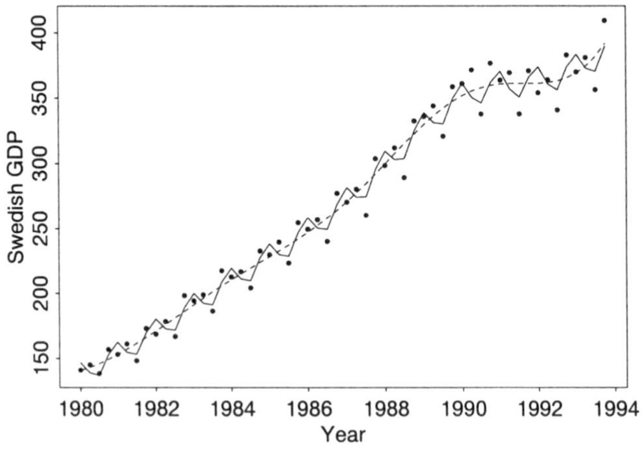

FIGURE 15.4. The gross domestic product for Sweden with seasonal variation. The solid line is the smooth using operator $L = (-\gamma D + D^2)(\omega^2 I + D^2)$, and the dashed line is the smooth for $L = D^4$, the smoothing parameter being determined by minimizing the GCV criterion in both cases.

seasonal variation. This suggests composing the operator $-\gamma D + D^2$ used for the U.S. GDP data with the deseasonalizing operator $\omega^2 I + D^2$ to obtain the composite operator of order four

$$L = (-\gamma D + D^2)(\omega^2 I + D^2) = -\gamma\omega^2 D + \omega^2 D^2 - \gamma D^3 + D^4. \quad (15.23)$$

This annihilates the four linearly independent functions given by the components of

$$u(t) = (1, \exp \gamma t, \sin \omega t, \cos \omega t)'.$$

In this application we know that $\omega = 2\pi$ for time measured in years, and the nonlinear least squares estimate for γ was 0.078.

The minimum GCV L-spline for these data is the solid line in the figure, and has $\text{GCV} = 142.9, \text{SSE} = 5298$, and $\text{DF}_1 = 10.4$. This fairly low-dimensional spline tracks both the seasonal and long-term variation rather well. By contrast, the minimum GCV polynomial spline corresponding to $L = D^4$ is shown by the dashed line, and has $\text{GCV} = 193.8, \text{SSE} = 8169$, and $\text{DF}_1 = 7.4$. As both the curve itself and the GCV value indicate, the polynomial spline was completely unable to model the seasonal variation, and treated it as noise. On the other hand, reducing the smoothing parameter λ to the point where SSE was reduced to the same value as was attained for the L-spline required $DF_1 = 28.2$, or nearly three times the degrees of freedom. Again we see that building the capacity to model important sources of variation into the operator L pays off handsomely.

15.5.4 Smoothing simulated human growth data

One of the triumphs of nonparametric regression techniques has been their capacity to uncover previously unsuspected aspects of growth in skeletal height (Gasser, Müller, *et al.* 1984; Ramsay, Bock and Gasser, 1995). In this illustration, spline smoothing with an estimated differential operator was applied to simulated smoothing data. The objective was to see whether estimating the smoothing operator improves the estimation of the growth curve itself, and of the acceleration function of stature, over an a priori 'off-the-rack' smoother.

To investigate how the performance of the L-spline would compare in practice with a polynomial spline, we simulated data to resemble actual human growth curve records. We generated two samples. A training sample of 100 records was analysed in a manner representative of actual practice, and a validation sample of 1000 records was used to see how these analyses would perform on data for which the analyses were not tuned.

The simulated data for both the training and validation samples consisted of growth records generated by using the triple logistic parametric nine-parameter growth model proposed by Bock and Thissen (1980). According to this model, height $h_i(t)$ at age t for individual i is

$$h_i(t) = \sum_{j=1}^{3} c_{ij}/[1 + \exp(-a_{ij}(t - b_{ij}))]. \tag{15.24}$$

Although not completely adequate to account for actual growth curves, this model does capture their salient features rather well. The actual number of parameters in the model turns out to be only eight, since the parameter a_{i1} can be expressed as a function of the other parameters.

We generated each record by first sampling from a population of coefficient vectors having a random distribution estimated from actual data for males in the Fels Growth Study (Roche, 1992). We computed the errorless growth curves (in cm) for the 41 age values ranging from 1 to 21 in half-yearly steps, and generated the simulated data by adding independent normal error with mean 0 and standard deviation 0.5 to

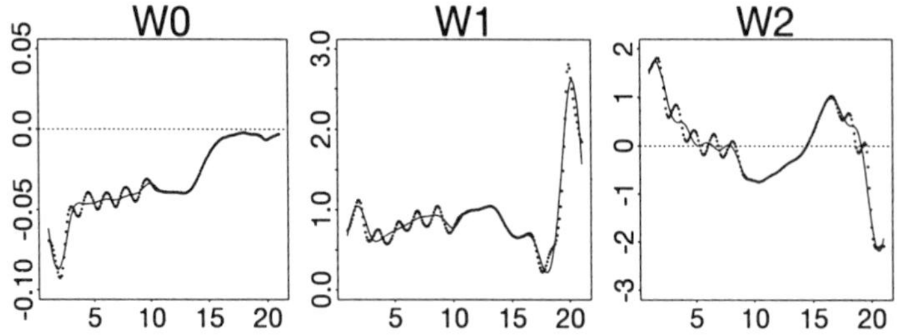

FIGURE 15.5. The three weight functions w_0, w_1, and w_2 for the operator $L = w_0 I + w_1 D + w_2 D^2 + D^3$. The points indicate the pointwise approximation, and the solid line indicates the basis function expansion.

these values. These simulated data had roughly the same variability as actual growth measurements.

The first step was to use the training sample to estimate the L-spline of third order that comes as near as possible to annihilating the curves. To this end, the first analysis consisted of polynomial spline smoothing of the simulated data to get estimates of the first three derivatives. The smoothing operator used for this purpose was D^5, implying that the smoothing splines were piecewise polynomials of degree nine. This permitted us to control the roughness of the third derivative in much the same way as a cubic smoothing spline controls the roughness of the smoothing function itself. The smoothing parameter was chosen to minimize the GCV criterion, and with this amount of replicated data, the value of its minimum is sharply defined. Since our principal differential analysis estimate of the operator L required numerical integration, we also obtained function and derivative estimates at 201 equally spaced values 1(.1)21.

We estimated a third-order differential operator L using both the pointwise technique and the basis expansion approach described in Chapter 14. For the latter approach, we used the 23 B-splines of order four defined by placing knots at the integer values of age. Figure 15.5 displays the estimated weight functions w_0, w_1, and w_2 for the operator $L = w_0 I + w_1 D + w_2 D^2 + D^3$. Although these are difficult to interpret, we can see that w_0 is close to 0, suggesting that the operator could be simplified by dropping the first term. On the other hand,

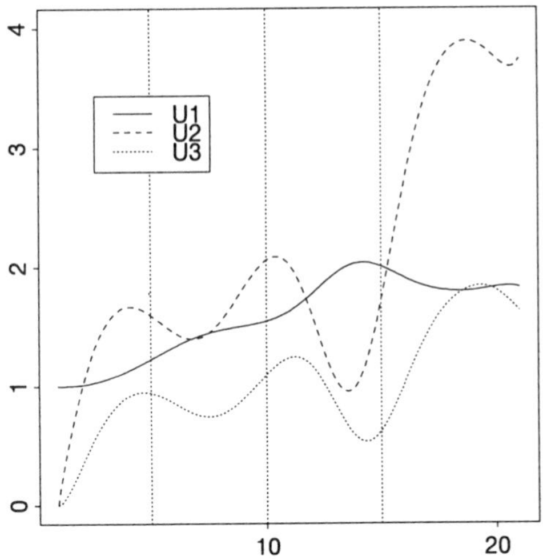

FIGURE 15.6. The three solutions to the homogeneous equation $Lu = 0$ corresponding to the linear differential operator L estimated for the simulated human growth data.

w_1 is close to one until age 15 when the growth function has strong curvature as the pubertal growth spurt ends, and its variation after age 15 helps the operator to deal with this. The acceleration weight w_2 varies substantially over the whole range of ages.

Figure 15.6 shows the three solutions u_j to $Lu = 0$, computed by Picard's method (13.22). Linear combinations of these three functions can produce good approximations to actual growth curves.

The next step was to use the estimated functions u_j and the techniques of Chapter 13 to estimate the Green's function G and the reproducing kernel k_2 associated with this operator. We approximated the integrals involved by applying the trapezoidal rule to the values at the 201 equally spaced arguments.

Now we were ready to carry out the actual smoothing of the training sample data by using the two techniques, L-spline and polynomial spline smoothing, both of order three, just as one would in practice. For both techniques, we relied on the GCV criterion to choose the smoothing parameter. The polynomial smooth gave values of GCV, DF_1 and λ of 487.9, 9.0 and 4.4, respectively, and the L-spline smooth produced corresponding values of 348.2, 11.2 and 0.63.

How well would these two smoothing techniques approximate the curves generating the data? To answer this question, we generated

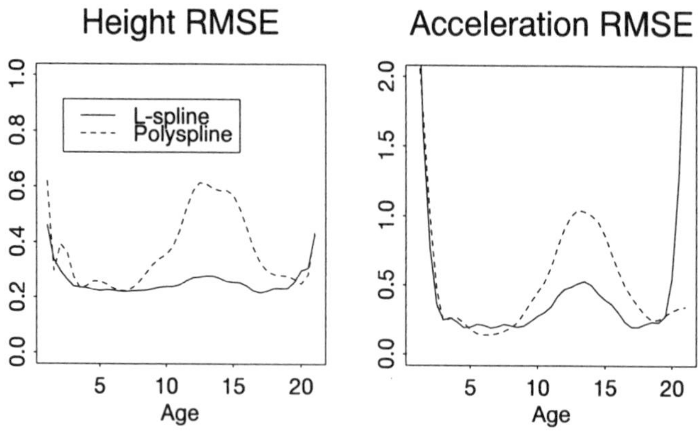

FIGURE 15.7. The left panel displays root-mean-squared error (RMSE) as a function of age for the simulated growth data. The solid line is for smoothing by using the estimated differential operator L, and the dashed line is for polynomial smoothing with $L = D^3$. The right panel shows these results for the estimated height acceleration.

1,000 new simulated curves using the same generation process, and applied the two smoothers using the training sample values of λ. Since we knew the values of the true curves, we could estimate the root-mean-squared error criterion

$$\text{RMSE}(t) = \sqrt{\mathrm{E}\{\hat{x}(t) - x(t)\}^2}$$

by averaging the squared error across the 1,000 curves for a given specific age t, and then taking the square root. This yielded the two RMSE curves displayed in Figure 15.7. We see that the estimate of both the growth curve itself and its acceleration by the L-spline procedure is much better for all but the final adult period, where the L-spline estimate of the acceleration curve becomes rather noisy and unstable. The improvement in the estimation of both curves is especially impressive prior to and during the pubertal growth spurt. The mean square error for the polynomial smooth is about four times that of the L-spline smooth, so using the L-spline is roughly equivalent to using the polynomial smooth with quadruple the sample size.

16
Some perspectives on FDA

16.1 The context of functional data analysis

We conclude this volume with some historical remarks and pointers to bibliographic references which have not been included in the main course of our development. Of course, we are acutely aware that many branches of statistical science consider functional models and the data that go with them. Functional data analysis has a long historical shadow, extending at least back to the attempts of Gauss and Legendre to estimate a comet's trajectory (Gauss, 1809; Legendre, 1805). So what we offer here is perhaps little more than a list of personal inspirations. In addition we suggest some directions for further research.

16.1.1 Replication and regularity

Although we want to leave the question of exactly what constitutes FDA soft around the edges, functional data problems as we have described them have two general features: replication and regularity. These are intimately related. Both permit the use of information from multiple data values to identify patterns; replication implies summaries taken across different observations, whereas regularity allows us to exploit information across argument values.

Replication is closely bound up with the key concept of a functional observation as a single entity, rather than a set of individual numbers or values. The availability of a sample of N related functional observations then leads to an interest in structure and variability in the data that requires more than one observation to detect. This is in contrast with much of the literature on nonparametric regression or curve estimation, where the focus is on estimating a single curve.

Functional principal components analysis, regression analysis, and canonical correlation, like their multivariate counterparts, characterize variation in terms of features with stability across replicates. Likewise, principal differential analysis and the use of an estimated linear differential operator for smoothing presume a model structure that belongs to the entire sample. Even curve registration aims to remove one important source of inter-curve variation so as to render more obvious the structure that remains.

Regularity implies that we exploit the smoothness of the processes generating the data, even though these data usually come to us in discrete form, and most of the analyses that we have considered assume a certain number of derivatives. The roughness penalty approach used throughout the book controls the size of derivatives or mixtures of derivatives of the estimated functional parameters. In this way we stabilize estimated principal components, regression functions, monotonic transformations, canonical weight functions, and linear differential operators.

Are there more general concepts of regularity that would aid FDA? For example, wavelet approaches to smoothing, briefly discussed in Section 3.2.6, are probably relevant, because of their ability to accommodate notions of regularity that, nevertheless, allow certain kinds of local misbehaviour.

Independent identically distributed observations are only one type of regularity. For example, can we use the replication principle implicit in stationary time series for the case where the values of the process are functions, to define useful functional data analyses? Besse and Cardot (1997) offer an interesting first step in this direction.

16.1.2 Some functional aspects elsewhere in statistics

Analysis of variance is often concerned with within-replication treatments. While an ANOVA design, as a rule, does not assume that these treatments correspond to variation over time or some other continuum, in practice this is often the case. Consequently texts on ANOVA such as Maxwell and Delaney (1990) pay much attention to topics that arise naturally when treatments correspond to events such as days, related spatial positions, and so on. Modifications allowing for a more complex correlational structure for the residuals, and the use of contrasts to make inferences about linear, quadratic, and other types of trend across treatments are examples.

Cognate to functional data analysis are fields such as longitudinal data analysis (Diggle et al., 1994), analysis of repeated measurements (Keselman and Keselman, 1993 and Lindsey, 1993) and growth curve analysis. Two classic papers that use principal components analysis to describe the modes of variation among replicated curves are Rao (1958) and Tucker (1958); Rao (1987) offers a summary of his and others' work on growth curves. Two more recent applications are Castro, Lawton and Sylvestre (1986) and Grambsch et al. (1995).

But these and the many other studies of curve structure do not give the regularity of the phenomena a primary role, giving the emphasis more to replication. Likewise, empirical Bayes, hierarchical linear model, or multilevel linear model approaches do treat functional data in principle, with the added feature of using prior information, but the nature of the prior structure tends to be multivariate rather than functional. Nevertheless, we expect that further research will show that the experience gained and the tools developed in these collateral disciplines can be put to good use in FDA.

16.1.3 *Uses of derivative information*

Two efforts stand out as path-breaking attempts to use derivative information in data analysis. The first of these is a series of papers on human growth data beginning with Largo et al. (1978) that focussed on the shape of the acceleration function. By careful and innovative use of smoothing techniques, spline and kernel, they were able to isolate a hitherto ignored phenomenon, the so-called mid-spurt, or hump in the acceleration curve that precedes the pubertal growth spurt and occurs at around age seven to eight in almost all children of either gender. These studies confirmed a principle that we have seen in many of our own functional data analyses: exogenous influences and other interesting events are often much more apparent in some order of derivative than in the original curves.

On a somewhat more technical note, the thesis by Besse (1979) and his subsequent papers (Besse and Ramsay, 1986; Besse, 1980 and 1988) moved the French data analytic school into a new realm involving data that take values in spaces of functions possessing a certain number of derivatives. Besse's discussion of PCA in the context of observations in Sobelev space was much inspired by Dauxois and Pousse (1976) and the functional analytic approaches to spline smoothing by Atteia

(1965). Ramsay and Dalzell (1991), who coined the term functional data analysis, extended this line of work to linear models.

16.1.4 Functional analytic treatments of statistical methods

One topic clearly within the scope of FDA is the description of statistical methods using functional analysis. For example, principal components analysis is a technique that lends itself naturally to many types of generalization. The notion of the eigenanalysis of a symmetric matrix was extended to integral operators with symmetric kernels in the last century, and the Karhunen-Loève decomposition of more general linear operators (Karhunen, 1947; Loève, 1945) is essentially the singular value decomposition in a wider context.

Parzen's papers (1961, 1963) are classics, and have had a great influence on the spline smoothing literature by calling attention to the important role played by the reproducing kernel. Grenander (1981) contributed further development, and Eaton (1983) provided a systematic coverage of multivariate analysis using inner product space notation. Stone (1987) also proposed a coordinate-free treatment, and Small and McLeish (1994) give a more recent discussion of Hilbert space methods in statistics.

Applied mathematicians and statisticians in France have been particularly active in recasting procedures originally developed in a conventional discrete or multivariate setting into a functional analytic notational framework. Deville (1974) considered the PCA of functional observations with values in $\mathcal{L}^2$. Cailliez and Pagès (1976) wrote an influential textbook on multivariate statistics that was both functional analytic in notation and coordinate-free in a geometrical sense. This was a courageous attempt to present advanced concepts to a mathematically unsophisticated readership, and it deserves to be better known. Dauxois and Pousse (1976) produced a comprehensive and sophisticated functional analytic exposition of PCA and CCA that unhappily remains in unpublished form.

Although the exercise of expressing the usual matrix treatments of multivariate methods in the more general language of functional analysis is intrinsically interesting to those with a taste for mathematical abstraction, it also defined directly the corresponding methods for infinite-dimensional or functional data. Some facility in functional analysis is a decided asset for certain aspects of research in FDA, as it already is in many other branches of applied mathematics.

16.2 Challenges for the future

We now turn to a few areas where there is clearly need for further research. These should be seen as a small selection of the wide range of topics that a functional data analytic outlook opens up.

16.2.1 Probability and inference

The presence of replication inevitably invites some consideration of random functions and probability distributions on function spaces. Of course, there is a large literature on stochastic processes and random functions, but because of our emphasis on data analysis we have not emphasized these topics in the present volume.

In passing, we note that functional observations can be random in a rather interesting variety of ways. We pointed out in Section 15.3 that the problem of spline smoothing is intimately related to the theory of stochastic processes defined by the nonhomogenous linear differential equation $Lx = f$ where L is a deterministic linear differential operator and f is white noise. Should we allow for some stochastic behaviour or nonlinearity in L? Is white noise always an appropriate model for f? Should we look more closely at the behaviour of an estimate of f in defining smoothing criteria, functional data analyses, and diagnostic analyses and displays, exploiting this estimate in ways analogous to our use of residuals in regression analysis? There is a large literature on such *stochastic differential equations*; see, for example Øksendal (1995). Though stochastic differential equations are of great current interest in financial mathematics, they have had relatively little impact on statistics more generally. This seems like a way to go.

We discussed the extension of classical inferential tools, such as the t-test or F-ratio, to the functional domain. We often need simulation to assess the significance of statistics once we move beyond the context of inference for a fixed argument value t. For a rather different approach to inference that incorporates both theoretical arguments and simulation, see Fan and Lin (1996).

Because of the infinite-dimensional nature of functional variation, the whole matter of extending conventional methods of inference—whether parametric or nonparametric, Bayesian or frequentist—is one that will require considerable thought before being well understood. We consider that there is much to do before functional data analysis will have an inferential basis as developed as that of multivariate statistics.

16.2.2 *Asymptotic results*

There is an impressive literature on the asymptotic and other theoretical properties of smoothing methods. Although some would argue that theoretical developments have not always had immediate practical interest or relevance, there are many examples clearly directed to practical concerns. For a recent paper in the smoothing literature that addresses the issue of the positive interaction between theoretical and practical research, see, for example, Donoho et al. (1995).

Some investigations of various asymptotic distributional aspects of FDA are Dauxois et al. (1982), Besse (1991), Pousse (1992), Leurgans et al. (1993), Pezzulli and Silverman (1993), Kneip (1994), Kneip and Engel (1995), and Silverman (1996). Nevertheless, theoretical aspects of FDA have not been researched in sufficient depth, and it is hoped that appropriate theoretical developments will feed back into advances in practical methodology.

16.2.3 *Multidimensional arguments*

Although we have touched on multivariate functions of a single argument t, coping with more than one dimension in the domain of our functions has been mainly beyond our scope. However, of course there is a rapidly growing number of fields where data are organized by space instead of or as well as time. For example, consider the great quantities of satellite and medical image data, where spatial dimensionality is two or three and temporal dependence is also of growing importance.

There is a large and growing literature on spatial data analysis; see, for example, Cressie (1991) and Ripley (1991). Likewise, smoothing over two or more dimensions of variation is a subject of active research (Scott, 1992). In particular, Wahba (1990) has pioneered the extension of regularization techniques to multivariate arguments. In principle, there is no conceptual difficulty in extending our own work on FDA to the case of multivariate arguments by using the roughness penalties relevant to tensor or thin-plate splines. Indeed, the paper by Hastie et al. (1995) reviewed in Section 12.7 uses roughness penalty methods to address a functional data analysis problem with a spatial argument. However, there are questions about multivariate roughness penalty methods in FDA that require further research.

16.2.4 *Practical methodology and applications*

Clearly, much research is needed on numerical methods, as is evident when one considers the work on something as basic as the pointwise linear model underlying spline smoothing. We think that regularization techniques will play a strong role, in part because they are so intuitively appealing. But of course there are often simpler approaches that may work more or less as well.

It is our hope that this book will give impetus to the wider dissemination and use of FDA techniques. More important than any of the detailed methodological issues raised in this chapter, the pressing need is for the widespread use of functional data analytic techniques in practice.

16.2.5 *Back to the data!*

Finally, we say simply that the data that we have analysed, and our colleagues who collected them, are ultimately responsible for our understanding of functional data analysis. If what this book describes is found to deserve a name for itself, it will be because, with each new set of functional data, we have discovered challenges and invitations to develop new methods. Statistics shows its finest aspects when exciting data find existing statistical technology not entirely satisfactory. It is this process that informs this book, and ensures that unforeseen adventures in research await us all.

Appendix
Some algebraic and functional techniques

This appendix covers various topics involving matrix decompositions, projections in vector and function spaces, and the constrained maximization of certain quadratic forms through the solution of appropriate eigenequations.

A.1 Matrix decompositions and generalized inverses

We describe two important matrix decompositions, the singular value decomposition and the QR decomposition. Both of these are standard techniques in numerical linear algebra, and can be carried out within packages such as S-PLUS and MATLAB (Mathworks, 1993). We do not give any details of the way the decompositions are computed; for these see, for example, Golub and Van Loan (1989) or the standard numerical linear algebra package LINPACK (Dongarra et al., 1979).

A.1.1 Singular value decompositions

Suppose $\mathbf{Z}$ is an $m \times n$ matrix. For many purposes it is useful to carry out a *singular value decomposition* (SVD) of $\mathbf{Z}$. This expresses $\mathbf{Z}$ as the product of three matrices

$$\mathbf{Z} = \mathbf{UDV}' \tag{A.1}$$

where, for some integer $q \le \min(m, n)$,

- $\mathbf{U}$ is $m \times q$ and $\mathbf{U}'\mathbf{U} = \mathbf{I}_q$, where $\mathbf{I}_q$ is the identity matrix of order q
- $\mathbf{D}$ is a $q \times q$ diagonal matrix with strictly positive elements on the diagonal
- $\mathbf{V}$ is $n \times q$ and $\mathbf{V}'\mathbf{V} = \mathbf{I}_q$.

The diagonal elements $d_1, d_2, \ldots, d_q$ of $\mathbf{D}$ are called the *singular values* of $\mathbf{Z}$, and the SVD can always be carried out in such a way that the diagonal elements $d_1, d_2, \ldots, d_q$ satisfy

$$d_1 \ge d_2 \ge \ldots \ge d_q > 0. \tag{A.2}$$

In this case, the number q is equal to the rank of the matrix $\mathbf{Z}$, i.e. the maximum number of linearly independent rows or columns of $\mathbf{Z}$.

In the special case where $\mathbf{Z}$ is square and symmetric, the requirement that the diagonal elements of $\mathbf{D}$ are necessarily positive is usually dropped, but the matrices $\mathbf{U}$ and $\mathbf{V}$ are chosen to be identical. Furthermore we may allow $q \ge \operatorname{rank} \mathbf{Z}$. Then the d_i include all the nonzero eigenvalues of $\mathbf{Z}$, together with some or all of the zero eigenvalues if there are any. We have

$$\mathbf{Z} = \mathbf{UDU}' \text{ with } \mathbf{U}'\mathbf{U} = \mathbf{I}. \tag{A.3}$$

In addition, if $\mathbf{Z}$ is positive semi-definite (so that $x'\mathbf{Z}x \ge 0$ for all vectors x) then

$$d_1 \ge d_2 \ge \ldots \ge d_q \ge 0. \tag{A.4}$$

A.1.2 Generalized inverses

Given any $m \times n$ matrix $\mathbf{Z}$, define a *generalized inverse* or *g-inverse* of $\mathbf{Z}$ as any $n \times m$ matrix $\mathbf{Z}^-$ such that

$$\mathbf{ZZ}^-\mathbf{Z} = \mathbf{Z}. \tag{A.5}$$

If $m = n$ and $\mathbf{Z}$ is an invertible matrix, then it follows from (A.5) that $\mathbf{Z}^{-1}$ is a g-inverse of $\mathbf{Z}$. Furthermore, by pre- and postmultiplying (A.5) by $\mathbf{Z}^{-1}$, we see that $\mathbf{Z}^{-1}$ is the *unique* g-inverse of $\mathbf{Z}$ in this case.

In the more general case, the matrix $\mathbf{Z}^-$ is not necessarily unique, but a particular g-inverse, called the *Moore–Penrose g-inverse* $\mathbf{Z}^+$, can always be calculated using the singular value decomposition (A.1) of the matrix $\mathbf{Z}$. Set

$$\mathbf{Z}^+ = \mathbf{VD}^{-1}\mathbf{U}'. \tag{A.6}$$

It is easy to check that $\mathbf{Z}^+$ is a g-inverse of $\mathbf{Z}$ and also that

$$\mathbf{Z}^+\mathbf{ZZ}^+ = \mathbf{Z}^+ \text{ and } \mathbf{ZZ}^+ = \mathbf{UU}'. \tag{A.7}$$

A.1.3 The QR decomposition

The *QR decomposition* of an $m \times n$ matrix $\mathbf{Z}$ is a different decomposition that yields the expression

$$\mathbf{Z} = \mathbf{QR}$$

where $\mathbf{Q}$ is an $m \times m$ orthogonal matrix (so that $\mathbf{Q}'\mathbf{Q} = \mathbf{QQ}' = \mathbf{I}$) and $\mathbf{R}$ is an $m \times n$ upper-triangular matrix (so that $\mathbf{R}_{ij} = 0$ if $i > j$).

If $m > n$ then the last $(m - n)$ rows of $\mathbf{R}$ will be zero, and each of the last $(m - n)$ columns x of $\mathbf{Q}$ will satisfy $x'\mathbf{Z} = 0$. Dropping these rows and columns will yield a *restricted QR decomposition* $\mathbf{Z} = \mathbf{Q}_1\mathbf{R}_1$ where $\mathbf{R}_1$ is an $n \times n$ upper-triangular matrix and $\mathbf{Q}$ is an $m \times n$ matrix of orthonormal columns.

A.2 Projections

In discussing the key concept of *projection*, first we consider projection matrices in m-dimensional spaces, and then go on to consider more general inner product spaces.

A.2.1 Projection matrices

Suppose that an $m \times m$ matrix $\mathbf{P}$ has the property that $\mathbf{P}^2 = \mathbf{P}$. Define $\mathcal{P}$ to be the subspace of R^m spanned by the columns of $\mathbf{P}$. The matrix $\mathbf{P}$ is called a *projection matrix* onto the subspace $\mathcal{P}$. The following two properties, which are easily checked, give the reason for this definition:

1. Every m-vector z is mapped by $\mathbf{P}$ into the subspace $\mathcal{P}$.
2. If z is already a linear combination of columns of $\mathbf{P}$, so that $z = \mathbf{P}u$ for some vector u, then $\mathbf{P}z = z$.

If $\mathbf{P}$ is a symmetric matrix, then $\mathbf{P}$ is called an *orthogonal* projection matrix, and will have several nice properties. For example, for any vector z we have

$$(\mathbf{P}z)'\{(\mathbf{I} - \mathbf{P})z\} = z'\mathbf{P}'(\mathbf{I} - \mathbf{P})z = z'(\mathbf{P}z - \mathbf{P}^2z) = 0.$$

This means that the projected vector $\mathbf{P}z$ and the 'residual vector' $z - \mathbf{P}z$ are orthogonal to one another, in the usual Euclidean sense. Furthermore, suppose that v is any vector in $\mathcal{P}$. Then, by a very similar argument,

$$v'(z - \mathbf{P}z) = (\mathbf{P}v)'(\mathbf{I} - \mathbf{P})z = v'\mathbf{P}(\mathbf{I} - \mathbf{P})z = 0,$$

so that the residual vector is orthogonal to all vectors in $\mathcal{P}$.

Suppose that w is any vector in $\mathcal{P}$ other than $\mathbf{P}z$. Then $w - \mathbf{P}z$ is also in $\mathcal{P}$ and therefore is orthogonal to $z - \mathbf{P}z$. Defining $\langle x, y \rangle = x'y$ and $\|x\|$ to be the usual Euclidean inner product and norm,

$$\begin{aligned} \|z - w\|^2 &= \|z - \mathbf{P}z\|^2 + \|\mathbf{P}z - w\|^2 + 2\langle z - \mathbf{P}z, \mathbf{P}z - w \rangle \\ &= \|z - \mathbf{P}z\|^2 + \|\mathbf{P}z - w\|^2 > \|z - \mathbf{P}z|^2. \end{aligned} \tag{A.8}$$

This means that $\mathbf{P}z$ is the point closest to z in the subspace $\mathcal{P}$. Thus orthogonal projections onto a subspace have the property of mapping each vector to the nearest point in the subspace.

More generally, if the inner product is $\langle x, y \rangle = x'\mathbf{W}y$, and if $\mathbf{P}$ is a projection onto the space $\mathcal{P}$ such that $\mathbf{WP}$ is symmetric, then $\mathbf{P}$ is orthogonal with respect to this inner product, meaning that $\langle \mathbf{P}z, z - \mathbf{P}z \rangle = 0$ and $\langle v, z - \mathbf{P}z \rangle = 0$ for all v in $\mathcal{P}$.

A.2.2 Finding an appropriate projection matrix

Now suppose we are not given a projection matrix, but instead we are given a subspace $\mathcal{U}$ of R^m, and we wish to find an orthogonal projection matrix $\mathbf{P}$ that projects onto $\mathcal{U}$.

Let $\mathbf{Z}$ be any matrix whose columns are m-vectors that span the subspace $\mathcal{U}$. There is no need for the columns to be linearly independent. Define $\mathbf{P}$ by

$$\mathbf{P} = \mathbf{Z}\mathbf{Z}^{-}.$$

It is straightforward to show that $\mathbf{P}$ is a projection onto the subspace $\mathcal{U}$ as required.

To get an *orthogonal* projection, define $\mathbf{P}$ using the Moore-Penrose g-inverse $\mathbf{Z}^{+}$. Then, in terms of the SVD of $\mathbf{Z}$, $\mathbf{P} = \mathbf{U}\mathbf{U}'$, so that $\mathbf{P}$ is a symmetric matrix and hence an orthogonal projection.

A.2.3 Projections in more general inner product spaces

We can extend these ideas to projections in more general inner product spaces as discussed in Section 2.4.1. As in that section, let $u_1, \ldots, u_n$ be any n elements of our space, and let u be the n-vector whose elements are $u_1, \ldots, u_n$. Let $\mathcal{U}$ be the subspace consisting of all possible linear combinations $c'u$ for real n-vectors c. Suppose that P is an orthogonal projection onto $\mathcal{U}$ as specified in Section 2.4.1. The proof that P maps each element z to the nearest member Pz of $\mathcal{U}$ is identical to the argument given in (A.8) because that depends only on the defining properties of an inner product and associated norm.

How are we to find an orthogonal projection of this kind? Extend our notation to define $\mathbf{K} = \langle u, u' \rangle$ as the symmetric $n \times n$ matrix with elements $\langle u_i, u_j \rangle$. The matrix $\mathbf{K}$ is positive semi-definite since $x'\mathbf{K}x = \langle x'u, u'x \rangle = \|x'u\|^2 \geq 0$ for any real n-vector x,

Define the operator P by

$$Pz = u'\mathbf{K}^+ \langle u, z \rangle$$

for all z. By definition Pz is a linear combination of the elements of u and hence is in $\mathcal{P}$. We shall show that P is an orthogonal projection onto $\mathcal{P}$.

If y is in $\mathcal{P}$, then $y = u'c$ for some real vector c, so that $Py = u'\mathbf{K}^+\mathbf{K}c$, and $y - Py = u'd$ where $d = (\mathbf{I} - \mathbf{K}^+\mathbf{K})c$. Therefore, since $\mathbf{K}\mathbf{K}^+\mathbf{K} = \mathbf{K}$,

$$\|y - Py\|^2 = d'\mathbf{K}d = d'(\mathbf{K} - \mathbf{K}\mathbf{K}^+\mathbf{K})c = 0,$$

implying that $\|y - Py\|^2 = 0$ and $Py = y$.

Finally, given any v in $\mathcal{P}$, and any z, use the property (A.7) to show that

$$\begin{aligned} \langle Pz - v, Pz \rangle &= \langle P(z - v), Pz \rangle = \langle z - v, u' \rangle \mathbf{K}^+\mathbf{K}\mathbf{K}^+ \langle u, z \rangle \\ &= \langle z - v, u' \rangle \mathbf{K}^+ \langle u, z \rangle = \langle Pz - v, z \rangle \end{aligned}$$

and therefore that $\langle Pz - v, z - Pz \rangle = 0$, completing the proof that P is the required orthogonal projection onto $\mathcal{P}$.

A.3 Constrained maximization of a quadratic function

A.3.1 *The finite-dimensional case*

Suppose that $\mathbf{A}$ is a symmetric $p \times p$ matrix. An important result in linear algebra concerns the constrained maximization problem

$$\max x'\mathbf{A}x \text{ for } p\text{-vectors } x \text{ subject to } x'x = 1. \tag{A.9}$$

Let $\lambda_1 \geq \lambda_2 \geq \ldots \geq \lambda_p$ be the eigenvalues of $\mathbf{A}$, and let u_i be the corresponding eigenvectors, each normalized to have $\|u_i\| = 1$. Let $\mathbf{U}$ be the matrix whose columns are the eigenvectors u_i and let $\mathbf{D}$ be the diagonal matrix with diagonal elements λ_i. Then $\mathbf{A} = \mathbf{UDU}'$, and $\mathbf{UU}' = \mathbf{U}'\mathbf{U} = \mathbf{I}$.

Set $y = \mathbf{U}'x$ in (A.9), so that $x = \mathbf{U}y$ and $x'x = y'\mathbf{U}'\mathbf{U}y = y'y$, so the constraint $x'x = 1$ is equivalent to $y'y = 1$. Therefore, in terms of y, the maximization problem (A.9) can be rewritten as

$$\max y'\mathbf{D}y \text{ for } p\text{-vectors } y \text{ subject to } y'y = 1. \tag{A.10}$$

This is clearly solved by setting y to be the vector $(1, 0, \ldots, 0)'$, so that x is the first column of $\mathbf{U}$, in other words the leading normalized eigenvector u_1 of $\mathbf{A}$.

By an extension of this argument, we can characterize all the eigenvectors of $\mathbf{A}$ as solutions of successive optimization problems. The jth normalized eigenvector u_j solves the problem (A.9) subject to the additional constraint of being orthogonal to all the solutions found so far:

$$\max x'\mathbf{A}x \text{ subject to } x'x = 1 \text{ and } x'u_1 = x'u_2 = \ldots = x'u_{j-1} = 0. \tag{A.11}$$

Setting $x = u_j$ gives $x'\mathbf{A}x = \lambda_j u_j' u_j = \lambda_j$, the jth eigenvalue.

A.3.2 The problem in a more general space

Now suppose we are working within a more general inner product space. The role of a symmetric matrix is now played by a self-adjoint linear operator A, that is one satisfying the condition

$$\langle x, Ay \rangle = \langle Ax, y \rangle \text{ for all } x \text{ and } y.$$

We shall assume that A is a completely continuous (or compact) symmetric transformation on a Hilbert space; there is no need at all for the reader to understand what this means, but anyone interested is referred to Riesz and Nagy (1956) or any other standard text on functional analysis. The reader can always take it on trust that the assumptions are satisfied when we appeal to the results of this section.

The problem

$$\max \langle x, Ax \rangle \text{ subject to } \|x\| = 1 \tag{A.12}$$

corresponds to the maximization problem (A.9), and we can define a sequence u_j as the solutions to the succession of optimization problems

$$\max \langle x, Ax \rangle \text{ subject to } \|x\| = 1 \text{ and } \langle x, u_i \rangle = 0 \text{ for } i < j. \tag{A.13}$$

Under the conditions referred to above, these optimization problems can be solved by considering the eigenfunction problem

$$Au = \lambda u$$

and normalizing the eigenfunctions u to satisfy $\|u\| = 1$. Suppose that the eigenvalues are $\lambda_1 \geq \lambda_2 \geq \ldots$ with eigenfunctions $u_1, u_2, \ldots$. Then the leading eigenfunction u_1 solves the optimization problem (A.12) and the value of the maximum is λ_1. The successive eigenfunctions u_j solve the constrained problem (A.13), and the maximum at the jth stage is $\langle u_j, Au_j \rangle = \lambda_j \|u_j\|^2 = \lambda_j$.

A.3.3 Generalized eigenproblems

Sometimes we wish to modify the optimization problems we have considered by introducing a positive-definite symmetric matrix $\mathbf{B}$ or self-adjoint operator B into the constraints, replacing the constraint $\|x\| = 1$ by $x'\mathbf{B}x = 1$ or, more generally, $\langle x, Bx \rangle = 1$, and similarly defining orthogonality with respect to the matrix $\mathbf{B}$ or operator B.

Consider the solutions of the *generalized eigenproblem*

$$Av = \rho Bv.$$

and normalize the solutions to satisfy $\langle v, Bv \rangle = 1$. Suppose that the solutions are $v_1, v_2, \ldots$, with corresponding generalized eigenvalues $\rho_1 \geq \rho_2 \geq \ldots$. Under suitable conditions, which are always satisfied in the finite-dimensional case, and are analogous to those noted above for more general spaces, the leading generalized eigenvector or function v_1 solves the problem

$$\max \langle v, Av \rangle \text{ subject to } \langle v, Bv \rangle = 1, \tag{A.14}$$

and the maximizing value is equal to ρ_1. The jth generalized eigenvector or function v_j solves the problem

$$\max \langle v, Av \rangle \text{ subject to } \langle v, Bv \rangle = 1 \text{ and } \langle v, Bv_i \rangle = 0 \text{ for } i < j$$

and the maximizing value is ρ_j.

Finally, we note that the problem of maximizing the ratio

$$\frac{\langle v, Av \rangle}{\langle v, Bv \rangle} \tag{A.15}$$

for $v \neq 0$ is equivalent to that of maximizing $\langle v, Av \rangle$ subject to the constraint $\langle v, Bv \rangle = 1$. To see this, note that scaling any v to satisfy the constraint does not affect the value of the ratio (A.15), and so the maximum of the ratio is unaffected by the imposition of the constraint. Once the constraint is imposed, the denominator of (A.15) is equal to 1, and so maximizing the ratio subject to $\langle v, Bv \rangle = 1$ is exactly the same as the original maximization problem (A.14).

References

Anderson, T. W. (1984) *An Introduction to Multivariate Statistical Analysis.* Second edition. New York: John Wiley & Sons.

Anselone, P. M. and Laurent, P. J. (1967) A general method for the construction of interpolating or smoothing spline-functions. *Numerische Mathematik,* **12**, 66-82.

Ansley, C. F. and Kohn, R. (1990) Filtering and smoothing in state space models with partially diffuse initial conditions. *Journal of Time Series Analysis,* **11**, 275-293.

Ansley, C. F., Kohn, R. and Wong, C.-M. (1993) Nonparametric spline regression with prior information. *Biometrika,* **80**, 75-88.

Atteia, M. (1965) Spline-fonctions généralisées. *Comptes Rendus de l'Académie des Sciences Série I: Mathématique,* **261**, 2149-2152.

Basilevsky, A. (1994) *Statistical Factor Analysis and Related Methods.* New York: John Wiley & Sons.

Besse, P. (1979) Etude descriptive des processus: Approximation et interpolation. Thèse de troisième cycle, Université Paul-Sabatier, Toulouse.

Besse, P. (1980) Deux exemples d'analyses en composantes principales filtrantes. *Statistique et Analyse des Données,* **3**, 5-15.

Besse, P. (1988) Spline functions and optimal metric in linear principal component analysis. In J. L. A. van Rijkevorsel and J. de Leeuw (eds) *Component and Correspondence Analysis: Dimensional Reduction by Functional Approximation.* New York: John Wiley & Sons.

Besse, P. (1991) Approximation spline de l'analyse en composantes principales d'une variable aléatoire hilbertienne. *Annales de la Faculté des Sciences de Toulouse,* **12,** 329-349.

Besse, P. and Ramsay, J. O. (1986) Principal components analysis of sampled functions. *Psychometrika,* **51,** 285-311.

Besse, P., Cardot, H. and Ferraty, F. (1997) Simultaneous non-parametric regressions of unbalanced longitudinal data. *Computational Statistics and Data Analysis,* to appear.

Besse, P. and Cardot, H. (1997) Approximation spline de la prévision d'un processus fonctionnel autorégressif d'ordre 1. *Canadian Journal of Statistics,* to appear.

Bock, R. D. and Thissen, D. (1980) Statistical problems of fitting individual growth curves. In F. E. Johnston, A. F. Roche and C. Susanne (eds) *Human Physical Growth and Maturation: Methodologies and Factors.* New York: Plenum.

Bookstein, F. L. (1991) *Morphometric Tools for Landmark Data: Geometry and Biology.* Cambridge: Cambridge University Press.

Buckheit, J. B, Olshen, R. A., Blouch, K. and Myers, B. D. (1997) Modeling of progressive glomerular injury in humans with lupus nephritis. *American Journal of Physiology,* to appear.

Buja, A., Hastie, T. and Tibshirani, R. (1989) Linear smoothers and additive models (with discussion). *Annals of Statistics,* **17,** 453-555.

Caillez, F. and Pagès, J. P. (1976) *Introduction à l'analyse des données.* Paris: SMASH.

Castro, P. E., Lawton, W. H. and Sylvestre, E. A. (1986) Principal modes of variation for processes with continuous sample curves. *Technometrics,* **28,** 329-337.

Char, B. W., Geeddes, K. O., Gonnet, G. H., Leong, B. L., Monagan, M. B. and Watt, S. M. (1991) *MAPLE V Language Reference Manual.* New York: Springer.

Chui, C. K. (1992) *An Introduction to Wavelets.* San Diego: Academic Press.

Cleveland, W. S. (1979) Robust locally weighted regression and smoothing scatterplots. *Journal of the American Statistical Association,* **74,** 829-836.

Coddington, E. A. (1989) *An Introduction to Ordinary Differential Equations.* New York: Dover.

Coddington, E. A. and Levinson, N. (1955) *Theory of Ordinary Differential Equations.* New York: McGraw-Hill.

Cook, R. D. and Weisberg, S. (1982) *Residuals and Influence in Regression.* New York: Chapman and Hall.

Craven, P. and Wahba, G. (1979) Smoothing noisy data with spline functions: estimating the correct degree of smoothing by the method of generalized cross-validation, *Numerische Mathematik*, **31**, 377–403.

Cressie, N. (1991) *Statistics for Spatial Data.* New York: John Wiley & Sons.

Dalzell, C. J. and Ramsay, J. O. (1990) Computing reproducing kernels with arbitrary boundary constraints. *SIAM Journal of Scientific Computing*, **14**, 511–518.

Daubechies, I. (1992) *Ten Lectures on Wavelets.* CBMS-NSF Series in Applied Mathematics, **61**. Philadelphia: Society for Industrial and Applied Mathematics.

Dauxois, J. and Pousse, A. (1976) Les analyses factorielles en calcul des probabilité et en statistique: Essai d'étude synthètique. Thèse d'état, Université Paul-Sabatier, Toulouse.

Dauxois, J., Pousse, A. and Romain, Y. (1982) Asymptotic theory for the principal component analysis of a vector random function: some applications to statistical inference. *Journal of Multivariate Analysis*, **12**, 136–154.

de Boor, C. (1978) *A Practical Guide to Splines.* New York: Springer.

Delves, L. M. and Mohamed, J. L. (1985) *Computational Methods for Integral Equations.* Cambridge: Cambridge University Press.

Deville, J. C. (1974) Méthodes statistiques et numériques de l'analyse harmonique. *Annales de l'INSEE*, **15**, 7–97.

Diggle, P. J., Liang, K.-Y. and Zeger, S. L. (1994) *Analysis of Longitudinal Data.* New York: Oxford University Press.

Dongarra, J. J., Bunch, J. R., Moler, C. B. and Stewart, G. W. (1979) *LINPACK Users' Guide.* Philadelphia: Society for Industrial and Applied Mathematics.

Donoho, D. L., Johnstone, I. M., Kerkyacharian, G. and Picard, D. (1995) Wavelet shrinkage: asymptopia? (with Discussion). *Journal of the Royal Statistical Society*, Series B, **57**, 301–369.

Eaton, M. L. (1983) *Multivariate Statistics: A Vector Space Approach.* New York: John Wiley & Sons.

Eaton, M. L. and Perlman, M. D. (1973) The non-singularity of generalized sample covariance matrices. *Annals of Statistics*, **1**, 710-717.

Engle, R. F., Granger, C. W. J., Rice, J. A. and Weiss, A. (1986) Semiparametric estimates of the relation between weather and electricity sales. *Journal of the American Statistical Association*, **81**, 310-320.

Eubank, R. L. (1988) *Spline Smoothing and Nonparametric Regression.* New York: Marcel Dekker.

Falkner, F. (ed.) (1960) *Child Development: An International Method of Study.* Basel: Karger.

Fan, J. and Gijbels, I. (1996) *Local Polynomial Modelling and its Applications.* London: Chapman and Hall.

Fan, J. and Lin, S.-K. (1996) Tests of significance when data are curves. Unpublished manuscript.

Friedman, J. H. and Silverman, B. W. (1989) Flexible parsimonious smoothing and additive modeling (with Discussion and Response). *Journal of the American Statistical Association*, **31**, 1-39.

Gasser, T. and Kneip, A. (1995). Searching for structure in curve samples. *Journal of the American Statistical Association*, **90**, 1179-1188.

Gasser, T., Kneip, A., Binding, A., Prader, A. and Molinari, L. (1991a) The dynamics of linear growth in distance, velocity and acceleration. *Annals of Human Biology*, **18**, 187-205.

Gasser, T., Kneip, A., Ziegler, P., Largo, R., Molinari, L., and Prader, A. (1991b) The dynamics of growth of height in distance, velocity and acceleration. *Annals of Human Biology*, **18**, 449-461.

Gasser, T., Kneip, A., Ziegler, P., Largo, R. and Prader, A. (1990) A method for determining the dynamics and intensity of average growth. *Annals of Human Biology*, **17**, 459-474.

Gasser, T. and Müller, H.-G., (1979) Kernel estimation of regression functions. In T. Gasser and M. Rosenblatt (eds) *Smoothing Techniques for Curve Estimation.* Heidelberg: Springer, pp. 23-68.

Gasser, T. and Müller, H.-G., (1984) Estimating regression functions and their derivatives by the kernel method. *Scandinavian Journal of Statistics*, **11**, 171-185.

Gasser, T. , Müller, H.-G., Köhler, W., Molinari, L. and Prader, A. (1984). Nonparametric regression analysis of growth curves. *Annals of Statistics*, **12**, 210-229.

Gauss, C. F. (1809) *Theoria motus corporum celestium*. Hamburg: Perthes et Besser.

Golub, G. and Van Loan, C. F. (1989) *Matrix Computations*. Second edition. Baltimore: Johns Hopkins University Press.

Grambsch, P. M., Randall, B. L. Bostick, R. M., Potter, J. D. and Louis, T. A. (1995) Modeling the labeling index distribution: an application of functional data analysis. *Journal of the Americal Statistical Association*, **90**, 813-821.

Green, P. J. and Silverman, B. W. (1994) *Nonparametric Regression and Generalized Linear Models: A Roughness Penalty Approach*. London: Chapman and Hall.

Greenhouse, S. W. and Geisser, S. (1959) On methods in the analysis of profile data. *Psychometrika*, **24**, 95-112.

Grenander, U. (1981) *Abstract Inference*. New York: John Wiley & Sons.

Härdle, W. (1990) *Applied Nonparametric Regression*. Cambridge: Cambridge University Press.

Hastie, T. and Tibshirani, R. (1990) *Generalized Additive Models*. New York: Chapman and Hall.

Hastie, T., Buja, A. and Tibshirani, R. (1995) Penalized discriminant analysis. *Annals of Statistics*, **23**, 73-102.

Heckman, N. and Ramsay, J. O (1996) Penalized regression with model-based penalties. University of British Columbia: unpublished manuscript.

Houghton, A. N., Flannery, J. and Viola, M. V. (1980) Malignant melanoma in Connecticut and Denmark. *International Journal of Cancer*, **25**, 95-104.

Hutchison, M. F. and de Hoog, F. R. (1985) Smoothing noisy data with spline functions. *Numerische Mathematik*, **47**, 99-106.

Huynh, H. S. and Feldt, L. (1976) Estimation of the Box correction for degrees of freedom from sample data in randomized block and split-plot designs. *Journal of Educational Statistics*, **1**, 69-82.

Ince, E. L. (1956) *Ordinary Differential Equations*. New York: Dover.

Johnson, R. A. and Wichern, D. A. (1988) *Applied Multivariate Statistical Analysis*. Englewood Cliffs, N. J.: Prentice Hall.

Johnstone, I. M. and Silverman, B. W. (1997) Wavelet threshold estimators for data with correlated noise. *Journal of the Royal Statistical Society, Series B*, **59**, 319-351.

Karhunen, K. (1947) Über linear Methoden in der Warscheinlichkeitsrechnung. *Annales Academiae Scientiorum Fennicae*, **37**, 1-79.

Keselman, H. J. and Keselman, J. C. (1993) Analysis of repeated measurements. In L. K. Edwards (ed.) *Applied Analysis of Variance in Behavioral Science*. New York: Marcel Dekker, pp. 105-145.

Kimeldorf, G. S. and Wahba, G. (1970) A correspondence between Bayesian estimation on stochastic processes and smoothing by splines. *Annals of Mathematical Statistics*, **41**, 495-502.

Kneip, A. (1994) Nonparametric estimation of common regressors for similar curve data. *Annals of Statistics*, **22**, 1386-1427.

Kneip, A. and Engel, J. (1995) Model estimation in nonlinear regression under shape invariance. *Annals of Statistics*, **23**, 551-570.

Kneip, A. and Gasser, T. (1988) Convergence and consistency results for self-modeling nonlinear regression. *Annals of Statistics*, **16**, 82-112.

Kneip, A. and Gasser, T. (1992) Statistical tools to analyze data representing a sample of curves. *Annals of Statistics*, **20**, 1266-1305.

Largo, R. H., Gasser, T., Prader, A., Stützle, W. and Huber, P. J. (1978) Analysis of the adolescent growth spurt using smoothing spline functions. *Annals of Human Biology*, **5**, 421-434.

Legendre, A. M. (1805) *Nouvelles méthodes pour la détermination des orbites des comètes*. Paris: Courcier.

Leurgans, S. E., Moyeed, R. A. and Silverman, B. W. (1993) Canonical correlation analysis when the data are curves. *Journal of the Royal Statistical Society*, Series B, **55**, 725-740.

Lindsey, J. K. (1993) *Models for Repeated Measurements*. New York: Oxford University Press.

Loève, M. (1945) Fonctions aléatoires de second ordre. *Comptes Rendus de l'Académie des Sciences, Série I: Mathématique*, **220**, 469.

MathWorks Inc. (1993) *MATLAB Reference Guide*. Natick, Mass.: The MathWorks Inc.

Maxwell, S. E. and Delaney, H. D. (1990) *Designing Experiments and Analyzing Data: A Model Comparison Perspective*. Belmont, Calif.: Wadsworth.

Mulaik, S. A. (1972) *The Foundations of Factor Analysis*. New York: McGraw-Hill.

Nadaraya, E. A. (1964) On estimating regression. *Theory of Probability and its Applications*, **10**, 186–190.

Nason, G. P. and Silverman, B. W. (1994) The discrete wavelet transform in S. *Journal of Computational and Graphical Statistics*, **3**, 163–191.

OECD (1995) *Quarterly National Accounts*, **3**.

Øksendal, B. (1995) *Stochastic Differential Equations: An Introduction with Applications*. New York: Springer.

Olshen, R. A., Biden, E. N., Wyatt, M. P. and Sutherland, D. H. (1989) Gait analysis and the bootstrap. *Annals of Statistics*, **17**, 1419–1440.

Parzen, E. (1961) An approach to time series analysis. *Annals of Mathematical Statistics*, **32**, 951–989.

Parzen, E. (1963) Probability density functionals and reproducing kernel Hilbert spaces, In M. Rosenblatt (ed.) *Proceedings of the Symposium on Time Series Analysis*. Providence, RI.: Brown University, pp. 155–169.

Pezzulli, S. D. and Silverman, B. W. (1993) Some properties of smoothed principal components analysis for functional data. *Computational Statistics*, **8**, 1–16.

Pousse, A. (1992) Etudes asymptotiques. In J.-J. Droesbeke, B. Fichet and P. Tassi (eds) *Modèles pour l'Analyse des Données Multidimensionelles*. Paris: Economica.

Press W. H., Teukolsky S. A., Vetterling W. T. and Flannery B. P. (1992) *Numerical Recipes in Fortran*. Second edition. Cambridge: Cambridge University Press.

Ramsay, J. O. (1982). When the data are functions. *Psychometrika*, **47**, 379–396.

Ramsay, J. O. (1989) The data analysis of vector-valued functions. In E. Diday (ed.) *Data Analysis, Learning Symbolic and Numeric Knowledge*. Commack, N. Y.: Nova Science Publishers.

Ramsay, J. O. (1996a) Principal differential analysis: data reduction by differential operators. *Journal of the Royal Statistical Society*, Series B, **58**, 495–508.

Ramsay, J. O. (1996b) Estimating smooth monotone functions. McGill University: unpublished manuscript.

Ramsay, J. O. (1996c) Pspline: An S-PLUS module for polynomial spline smoothing. Computer software in the `statlib` archive.

Ramsay, J. O., Altman, N. and Bock, R. D. (1994) Variation in height acceleration in the Fels growth data. *Canadian Journal of Statistics*, **22**, 89-102.

Ramsay, J. O., Bock. R. D. and Gasser, T. (1995) Comparison of height acceleration curves in the Fels, Zurich, and Berkeley growth data. *Annals of Human Biology*, **22**, 413-426.

Ramsay, J. O. and Dalzell, C. J. (1991) Some tools for functional data analysis (with Discussion). *Journal of the Royal Statistical Society*, Series B, **53**, 539-572.

Ramsay, J. O, Heckman, N. and Silverman, B. W. (1997) Some general theory for spline smoothing. *Behavioral Research: Instrumentation, Methods, and Computing*, to appear.

Ramsay, J. O. and Li, X. (1996) Curve registration. McGill University, unpublished manuscript.

Ramsay, J. O., Munhall, K., Gracco, V. L. and Ostry, D. J. (1996) Functional data analyses of lip motion. *Journal of the Acoustical Society of America*, **99**, 3718-3727.

Ramsay, J. O., Wang, X. and Flanagan, R. (1995) A functional data analysis of the pinch force of human fingers. *Applied Statistics*, **44**, 17-30.

Rao, C. R. (1958) Some statistical methods for comparison of growth curves. *Biometrics*, **14**, 1-17.

Rao, C. R. (1987) Prediction in growth curve models (with discussion). *Statistical Science*, **2**, 434-471.

Reinsch, C. (1967) Smoothing by spline functions. *Numerische Mathematik*, **10**, 177-183.

Reinsch, C. (1970) Smoothing by spline functions II. *Numerische Mathematik*, **16**, 451-454.

Rice, J. A. and Silverman, B. W. (1991) Estimating the mean and covariance structure nonparametrically when the data are curves. *Journal of the Royal Statistical Society*, Series B, **53**, 233-243.

Riesz, F. and Nagy, B. Sz.- (1956) *Functional Analysis*. (Translated by L. F. Boron.) London: Blackie.

Ripley, B. D. (1991) *Statistical Inference for Spatial Processes*. Cambridge: Cambridge University Press.

Roach, G. F. (1982) *Green's Functions*. Second edition. Cambridge: Cambridge University Press.

Roche, A. (1992) *Growth, Maturation and Body Composition: The Fels Longitudinal Study 1929-1991.* Cambridge: Cambridge Press.

Scott, D. W. (1992) *Multivariate Density Estimation.* New York: John Wiley & Sons.

Schumaker, L. (1981) *Spline Functions: Basic Theory.* New York: John Wiley & Sons.

Seber, G. A. F. (1984) *Multivariate Observations.* New York: John Wiley & Sons.

Silverman, B. W. (1985) Some aspects of the spline smoothing approach to non-parametric regression curve fitting. *Journal of the Royal Statistical Society,* Series B, **47**, 1–52.

Silverman, B. W. (1994) Function estimation and functional data analysis. In A. Joseph, F. Mignot, F. Murat, B. Prum and R. Rentschler (eds) *First European Congress of Mathematics.* Basel: Birkhäuser. Vol II, pp 407–427.

Silverman, B. W. (1995). Incorporating parametric effects into functional principal components analysis. *Journal of the Royal Statistical Society,* Series B, **57**, 673–689.

Silverman, B. W. (1996) Smoothed functional principal components analysis by choice of norm. *Annals of Statistics,* **24**, 1–24.

Simonoff, J. S. (1996) *Smoothing Methods in Statistics.* New York: Springer.

Small, C. G. and McLeish, D. L. (1994) *Hilbert Space Methods in Probability and Statistical Inference.* New York: John Wiley & Sons.

Statistical Sciences (1995) *S-PLUS Guide to Statistical and Mathematical Analysis, Version 3.3.* Seattle: StatSci, a division of MathSoft, Inc.

Stoer, J. and Bulirsch, R. (1980) *Introduction to Numerical Analysis.* New York: Springer.

Stone, M. (1987) *Coordinate-Free Multivariable Statistics.* Oxford: Clarendon Press.

Tenenbaum, M. and Pollard, H. (1963) *Ordinary Differential Equations.* New York: Harper and Row.

Tucker, L. R. (1958) Determination of parameters of a functional relationship by factor analysis. *Psychometrika,* **23**, 19–23.

Vinod, H. D. (1976) Canonical ridge and econometrics of joint production. *Journal of Econometrics,* **4**, 147–166.

Wahba, G. (1978) Improper priors, spline smoothing and the problem of guarding against model errors in regression. *Journal of the Royal Statistical Society*, Series B, **40**, 364–372.

Wahba, G. (1990) *Spline Models for Observational Data*. Philadelphia: Society for Industrial and Applied Mathematics.

Wand, M. P. and Jones, M. C. (1995) *Kernel Smoothing*. London: Chapman and Hall.

Watson, G. S. (1964) Smooth regression analysis. *Sankhyā*, Series A, **26**, 101–116.

Weinert, H. L., Byrd, R. H. and Sidhu, G. S. (1980) A stochastic framework for recursive computation of spline functions: Part II, smoothing splines. *Journal of Optimization Theory and Applications*, **2**, 255–268.

Wilson, A. M., Seelig, T. J., Shield, R. A. and Silverman, B. W. (1996) The effect of imposed foot imbalance on point of force application in the equine. Technical report, Department of Veterinary Basic Sciences, Royal Veterinary College.

Wolfram, S. (1991) *Mathematica: A System for Doing Mathematics by Computer*. Second edition. Reading, Mass.: Addison-Wesley.

Index

Springer Series in Statistics

(continued from p. ii)

Pollard: Convergence of Stochastic Processes.
Pratt/Gibbons: Concepts of Nonparametric Theory.
Ramsay/Silverman: Functional Data Analysis.
Read/Cressie: Goodness-of-Fit Statistics for Discrete Multivariate Data.
Reinsel: Elements of Multivariate Time Series Analysis, 2nd edition.
Reiss: A Course on Point Processes.
Reiss: Approximate Distributions of Order Statistics: With Applications to Non-parametric Statistics.
Rieder: Robust Asymptotic Statistics.
Rosenbaum: Observational Studies.
Ross: Nonlinear Estimation.
Sachs: Applied Statistics: A Handbook of Techniques, 2nd edition.
Särndal/Swensson/Wretman: Model Assisted Survey Sampling.
Schervish: Theory of Statistics.
Seneta: Non-Negative Matrices and Markov Chains, 2nd edition.
Shao/Tu: The Jackknife and Bootstrap.
Siegmund: Sequential Analysis: Tests and Confidence Intervals.
Simonoff: Smoothing Methods in Statistics.
Small: The Statistical Theory of Shape.
Tanner: Tools for Statistical Inference: Methods for the Exploration of Posterior Distributions and Likelihood Functions, 3rd edition.
Tong: The Multivariate Normal Distribution.
van der Vaart/Wellner: Weak Convergence and Empirical Processes: With Applications to Statistics.
Vapnik: Estimation of Dependences Based on Empirical Data.
Weerahandi: Exact Statistical Methods for Data Analysis.
West/Harrison: Bayesian Forecasting and Dynamic Models, 2nd edition.
Wolter: Introduction to Variance Estimation.
Yaglom: Correlation Theory of Stationary and Related Random Functions I: Basic Results.
Yaglom: Correlation Theory of Stationary and Related Random Functions II: Supplementary Notes and References.